A Naturalis

Insects

of Britain & Northern Europe

ROBERT READ

PHOTOGRAPHIC CONSULTANT: PAUL STERRY

JOHN BEAUFOY PUBLISHING

First published in the United Kingdom in 2015 by John Beaufoy Publishing,
11 Blenheim Court, 316 Woodstock Road, Oxford OX2 7NS, England
www.johnbeaufoy.com

10 9 8 7 6 5 4 3 2 1

Photo Credits

Front cover: Common Carder Bee © Paul Sterry. **Back cover:** Sloe Shield Bug © Paul Sterry.

Title page: Stag Beetle © Paul Sterry. **Contents page**: 7 Spot Ladybird © Paul Sterry.
Main descriptions: photographs supplied by Nature Photographers Ltd (www.naturephotographers.co.uk).
All images were taken by Paul Sterry with the exception of the following: p.7 farmland, Robert Read; p.99
Common Hawker, Michael Hammett; p.101 Emperor Dragonfly, Roger Tidman; p.102 Four-spotted Chaser,
Rob Read; p.104 Keeled Skimmer, Richard Revels; p.104 Common Darter, Richard Revels; p.105 Black Darter,
Richard Revels; p.108 Azure Damselfly, Andrew Merrick; p.108 Common Blue Damselfly, Richard Revels; p.110
Small Red Damselfly, Richard Revels; p.114 Capsid Bug, Andy Callow; p.123 Mosquito, Geoff du Feu; p.124
Empid Fly, Andy Callow; p.125 Band-eyed Brown Horse Fly, S.C. Bisserot; p.126 Empid Fly, Andy Callow;
p.127 Snipe Fly, Andy Callow; p.128 *Helophilus pendulus*, Richard Revels; p.130 Greenbottle, Robert Read;
p.132 Lesser House-fly, Geoff du Feu; p.136 Leaf-cutter Bee, Richard Revels.

Great care has been taken to maintain the accuracy of the information contained in this work. However,
neither the publishers nor the author can be held responsible for any consequences arising from the use of the
information contained therein.

ISBN 978-1-909612-36-5

Edited, designed and typeset by Gulmohur Press

Printed and bound in Malaysia by Times Offset (M) Sdn. Bhd.

CONTENTS

Acknowledgements

There are many individuals and organizations that have assisted either directly or indirectly with the preparation of this book; they include numerous people who have supplied specimens to photograph or information on where to find them. Special thanks go to Paul Sterry for his advice, constant direction and help with picture selection, and Andrew Cleave, MBE for his invaluable support and assistance.

Area covered

This book describes species that can be encountered in Britain and Ireland, and neighbouring regions of mainland Europe, including northern and central France, north-west Germany, Belgium, Denmark, Holland, Sweden and Norway.

This large region encompasses a huge variety of habitat types and supports a diverse range of insects that number many thousands of species. The most common, widespread and eye-catching examples to be encountered are the subject of this book. Many are generalist species that have adapted to survive in a variety of conditions and habitat types, and can be found commonly across the region; some are specialists that are only associated with particular habitats and climates; others are migratory species that are only encountered at certain times of year and which occur in varying numbers annually.

Insects by Habitat

The region covered by this book embraces a wide range of habitats. Soil type, altitude, climate and the influence of man's activities all play a part in determining the diversity of plant species found in any given habitat. In turn, these have a profound bearing on insect species and numbers: to a greater or lesser degree, insect populations are directly linked to the plant species present in any given location; these provide food, shelter and protection during larval and pupal development, and are important for adults too, of course.

Each habitat harbours its own niches and specialist insect species have evolved to exploit these areas. Numerous examples can be cited, including: decaying wood in the forests supporting beetle populations; the dung dropped by grazing animals providing food for Dung Flies and Dor Beetles; and clean swift-flowing water, a vital habitat for caddis larvae to flourish. The following sections describe the most important habitat types for insects in Britain and northern Europe. Not all European habitats are included here and it is worth noting that in any given location there is likely to be some overlap between habitats, with elements present of more than one of those described.

GRASSLAND

In northern Europe, most grassland habitats have been created by man, primarily by the removal and clearing of vast areas of once-mature woodland. Starting with bare ground, grassland is the first stage of vegetative succession, the appearance of low-growing grasses beginning the recolonization process. If left unmanaged, these grassland areas would

Summer grassland

advance to the next stages through the appearance of a succession of herbaceous plants and eventually shrubs. Grazing and regular cutting prevents other species from becoming established, therefore maintaining the grassland habitat. As with other habitats, soil type, drainage, latitude and altitude in any one location all have an effect on the range and diversity of the flora and, consequently, the insect population.

The degree and sensitivity of grassland habitat management has a profound impact on the number and diversity of the insect species that it supports. Most insects rely on plants for food and protection at some point in their life-cycles. Allow an area to be overgrazed, or cut too aggressively and there will be consequences: for example, a butterfly larval food plant might be eliminated or the plant stem that a skipper butterfly requires to protect it during pupation could be destroyed. The result can be the local extinction of a species. But too little management can be just as damaging as too much.

WOODLAND AND FORESTS

Defined as an area that is populated with closely spaced mature trees, woodland and forests are traditionally seen as the final stage in habitat succession. A vast range of micro-habitat types occurs within these areas of woodland.

An ancient and unmanaged deciduous woodland will contain a wide selection of tree species and a specialist range of ground flora suited to the sheltered conditions the woodland provides. Dead trees and fallen timber will be left to decay, the nutrients recycled and returned to the soil. These factors ensure an ecosystem that supports a diverse range of insect species, ranging from the specialist beetles and flies that rely on rotting wood, to the moths and butterflies that require a particular food plant for larval development. Contrast this with a dense, managed coniferous plantation: typically a monoculture comprising a single tree species. With little or no light for ground flora to become established and a lack of decaying timber, the number and diversity of the insects that it supports will be severely restricted.

Downy Birch woodland, Scotland

Reed bed

WETLANDS AND MARSHES

Wetlands and marshes are locations that contain waterlogged ground, usually in the vicinity of a body of still or running water; all these sites can be excellent for the aquatic developmental stages of specialist insects. The heading embraces a wide range of habitats including pools, lakes, gravel pits, streams, rivers, canals, fens, ditches, bogs and coastal marshes. These support an equally wide range of flora from mature trees to sedges and pond weeds, attracting a range of specialist insects. Some of the insect species described in the following pages have aquatic stages in their life-cycles. However, there are many other species that are purely aquatic and these are beyond the scope of this book.

HEATHLAND AND MOORLAND

Heaths and moors are areas containing a specialist group of hardy plants suited to free-draining, acid soils. Heathland occurs in lowland areas; moorland is associated with harsher upland environments. Both contain a range of flora that has evolved to cope with acid soils, waterlogged soils and, in the case

Coldharbour Moor, Peak District

of upland areas, harsher weather generally. Species of heather and gorse prevail, attracting similarly hardy specialist insects. The tough, wiry nature of the flora and the greater prevalence of evergreen species extend the feeding season for some species of insects.

FARMLAND

Across swathes of northern Europe vast areas of deciduous woodland and natural flora have been removed for the intensive cultivation of crop plants. This loss of habitat in itself has a huge and negative impact on insect diversity and abundance. Added to this is the fact that some insect species are perceived as pests and treated as such, so that agricultural land is not generally the place to look for a healthy and diverse range of insect species. The routine spraying of insecticides has obvious catastrophic effects on insect populations; certain targeted species, such as the Cockchafer, have suffered virtual extinction in some areas in the past. On the plus side, various government and EU schemes encouraging the retention of hedgerows and the establishment of wide field margins have helped maintain some natural habitats for insects. The retention of some woodland, hedgerows and areas of scrub in locations inaccessible to farm machinery also ensures a degree of mixed habitat capable of supporting a varied insect population.

Hampshire farmland

HEDGEROWS AND WAYSIDES

Overgrown hedgerow and scrub

Although planted by man, many hedgerows are, in fact, quite ancient, some being established boundaries for centuries. Their managed nature prevents them from developing a complete woodland ecosystem, and in terms of succession they straddle the line between scrub (the first regenerational stage after grassland) and woodland. Many replicate a woodland edge and have species that are associated with this habitat.

Other environments are interfered with by man to a greater or lesser degree. A consequence of this is that many areas show various intermediate stages of vegetative regeneration, a good example being wasteland such as a disused railway line or derelict industrial site. 'Wayside' refers to some of these intermediate environments.

GARDENS

Gardens by their very nature are managed habitats that often include every stage of habitat regeneration, from the managed grassland of the lawn and low-growing shrubs akin to hedgerows, heaths and moorland to wetlands in the form of garden ponds and pockets of mature woodland. Certain plant species are commonly grown in greater numbers than would naturally occur, therefore attracting larger concentrations of insects that specialize in those plants. A good example of this phenomenon is the Rose Aphid and their often unwelcome summer infestations of the garden roses. By replicating a wide range of habitat types, a good proportion of the insects covered in this book – naturally allied to other specialist habitats – can and do occur in the garden.

A typical English garden

WHAT ARE INSECTS?

An adult insect has a body divided into three clearly defined segments: the head containing and supporting all the sensory organs and mouthparts; the thorax containing the muscle mass and from which the paired wings (absent in some species) and six legs are appended; and the abdomen which contains the main organs and reproductive parts.

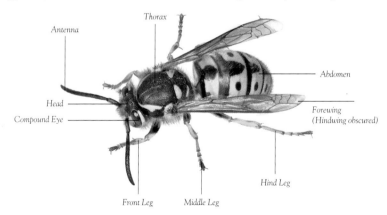

German Wasp, with body parts labelled

▪ What are Insects? ▪

To help understand and study the natural world, biologists categorize living organisms using a hierarchical taxonomic system. A knowledge of this system will assist in understanding the manner in which this book has been ordered.

Insects (Insecta) are classified as invertebrates, a section of the animal kingdom (Animalia) defined as animals without a backbone. They are contained within a subdivision (Phylum) of this group known as Arthropods, animals with an external skeleton, or exoskeleton, and a segmented body with paired and jointed limbs, the muscles being contained internally. Other unrelated members (Classes) of this subdivision are crustaceans and arachnids.

Classes are further divided into Orders, grouping together Species with similar characteristics. Not all of the insect Orders are mentioned below, but the following is a list of the major insect Orders represented by Species described in this book:

Lepidoptera	butterflies and moths
Thysanura	bristletails
Ephemeroptera	mayflies
Plecoptera	stoneflies
Orthoptera	grasshoppers, crickets and bush-crickets
Odonata	dragonflies and damselflies
Hemiptera	true bugs
Megaloptera	alderflies
Dermaptera	earwigs
Dictyoptera	cockroaches
Neuroptera	lacewings and allies
Mecoptera	scorpion flies
Raphidioptera	snake flies
Trichoptera	caddis flies
Diptera	true flies
Hymenoptera	bees, wasps, ants and allies
Coleoptera	beetles

Orders contain subgroups of related creatures called Families which themselves are further divided into Genera containing closely related Species; the Genera name forms the first part of the scientific name and second part is the specific name. By taking the Swallowtail Butterfly (*Papilio machaon*) as an example, the journey through its taxonomic classification can be followed:

Kingdom – Animalia

    Phylum – Arthropoda

        Class – Insecta

            Order – Lepidoptera

                Family – Papilionidae

                    Genus – *Papilio*

                        Species – *machaon*

Although common names for Species can differ greatly according to language and location, the scientific name of a species is accepted universally in all languages and countries.

STUDYING INSECTS

All of the species described in this book can be identified with no more than a small hand lens or digital camera. Modern digital cameras, even those that are now commonly found on the modern generation of mobile phones, often have excellent macro modes capable of taking high resolution images. These can be enlarged on the screen displaying more detail than can often be seen through a good hand lens; they are superb tools in aiding identification.

Digital cameras are a useful aid to species identification

For the budding entomologist keen on studying specimens in more detail, or discovering those that are active at night, varying methods of capture can be employed.

A traditional butterfly net is an effective tool for catching flying insects. Frames should be light and sturdy, the mesh fine and tough enough to prevent it from being ripped on brambles. Care should be taken with specimens captured in this manner as the wings are easily damaged irreparably. As well as flying insects, a more substantial net can be employed to 'sweep' for resting insects. By moving the net swiftly through low-growing vegetation, insects hiding on these plants may be collected in the mesh.

Sweeping for insects with a sturdy net

Shaking or striking tree branches and taller vegetation, while holding a tray or upturned umbrella underneath to catch falling objects, is a method known as 'beating'. It will often dislodge insects and larvae from their hiding places. Care should be exercised not to cause damage to vegetation.

Meticulous searching through shrubs and trees and on the ground underneath can reveal all sorts of interesting specimens from ground-crawling beetles to moths and shield bugs. Some species are too small or delicate to pick up by hand without causing damage. These can be collected with the aid of a

A moth trap in action

'pooter', a simple device comprising a collecting receptacle attached to a length of suction pipe, enabling the insect to be 'vacuumed up' unharmed.

Night-flying insects, and in particular moths, are often attracted by light. Employing a light trap to capture these species can be a rewarding exercise and one which is regularly used by moth enthusiasts. Equipment is readily available to purchase through specialist suppliers, but can prove costly. Plenty of advice on building your own is available by employing a little research. Alternatively, a single bright light over a white sheet works well to attract insects that can be caught with a butterfly net.

Many of the larger species of insect can be readily identified in the field without the need to capture them. Conversely, there are many smaller species that are not possible to accurately name beyond the identification of the Family group without studying them in detail under a microscope, often under dissection. These smaller species fall outside the scope of this book, but inevitably will be encountered in the field.

INSECT LIFE-CYCLES

Eggs

Larva

Pupa

Adult

■ CONSERVATION OF INSECTS ■

The vast majority of insects start their life as an egg laid by the adult; the egg is surrounded by a tough waterproof shell designed to survive harsh conditions. Many species lay eggs in summer or autumn and they have to endure the rigours of the winter months attached to leaves, buried in soil and decaying wood or submerged in shallow pools. Once hatched, the insect enters the growing stage of its life-cycle and this is referred to as a larva. During its growth, the larva has to moult its skin a number of times; each stage between moults is referred to as an 'instar'. The life-cycle then progresses in one of two ways.

INCOMPLETE METAMORPHOSIS

The larva becomes progressively more adult-like with each instar. Previously, the larval stages of insects that develop through incomplete metamorphosis were referred to as 'nymphs'.

COMPLETE METAMORPHOSIS

Keeping its form, the larva gets larger with each instar until fully grown when it metamorphoses into a pupa. A pupa is a hard-cased stage in the life-cycle that encloses and protects the internal organs and body parts of the insect as it goes through the final stages of metamorphosis. As with the egg, pupae provide protection from adverse conditions, many species overwintering in this phase. The insect emerges as a fully formed adult once the metamorphosis is complete.

CONSERVATION OF INSECTS

Although overlooked by some people or dismissed as insignificant, insects are unrivalled in the animal world in terms of sheer numbers and variety of species. They have adapted to survive and colonize every terrestrial and freshwater habitat on the planet outside the extreme Polar Regions. They often form immense colonies that number in the tens of thousands and are a vital biomass in ecosystems from a global to a micro level. Remove the habitat that supports them, control their numbers with pesticides or non-native predatory species, and the effects will permeate up the food chain.

Vast areas of natural habitat are outside the control of most of us as individuals. But areas that we do influence include our gardens, whether they are restricted to small window-boxes or encompass many acres of land. Active garden management and an understanding of individual habitat requirements can result in a balanced and diverse ecosystem and even encourage a particular target species. For example, leave an area of lawn uncut or a flower border untended for a couple of years and just see how many species of butterfly your garden will support. By controlling the environment within the garden, dynamic mini nature reserves can be created and form important and diverse insect concentrations in certain areas.

Attitudes towards the collection and recording of species have changed in recent years and many species of insect are now protected by law because of declining numbers. For decades the killing and pinning of specimens in display cases was accepted as standard practice. Thankfully, times have changed and this rather Victorian method of record-keeping is being replaced by modern digital photography. The need for expensive specialist camera equipment has gone. Through camera phones or low-budget cameras, vast collections of images can be quickly amassed and shared with the many specialist entomology groups that are springing up via social networking and photo sharing websites.

GLOSSARY

Common Wasp with head parts annotated

Abdomen The rear section of an insect's body usually appears segmented.

Antennae A pair of slender sensory organs projecting from the head.

Aquatic Living in water.

Arthropod An invertebrate with a hard exoskeleton and jointed legs.

Bioindicator A species used to monitor the impact of changes to an environment or ecosystem.

Bog Wetland habitat where the soil and water are acid.

Calcareous Calcium-based soil type, usually chalk or limestone.

Caterpillar The larva of a butterfly or moth.

Cerci Paired appendages at the tail end of the abdomen.

Chrysalis see 'Pupa'.

Compound eye An eye made up of a series of cells and lenses.

Coniferous Tree that bears its reproductive elements in cones.

Deciduous Tree that sheds its leaves in the autumn.

Diurnal Active during the day.

Dorsal Upper surface.

Double-brooded Completing two life-cycles in a single season.

Downland Grassland on chalky soils.

Dun Sub-adult stage of a mayfly, describing its colouration.

Elytra Hardened forewings/wing cases of a beetle.

Emergent Plant with its base and roots in water, the remainder above the surface.

Exoskeleton An exterior skeleton.

Heathland Area of low vegetation on acid soils supporting heathers and other members of the heath family of plants.

Hibernation State of dormancy during the winter months when metabolism is greatly reduced.

Introduced Not native to a location.

Invertebrate Animal lacking a backbone.

Keel Narrow ridge to the underside of the body.

Larva Pre-adult growing stage of an insect.

Mandibles Paired jaw-like mouth-parts used for gripping, cutting and biting.

Metamorphosis Process by which an insect changes morphologically from one immature stage to another, or from an immature stage into an adult.

Migrant A species that spends its life in more than one place, often travelling large distances.

Moorland Grazed, upland areas of low vegetation on acid soils.

Nocturnal Active after dark.

Nymph Immature growing stage in some insects that becomes progressively more adult-like with each moult.

Ovipositor Egg-laying body part found in some female insects at the tail end of the abdomen.

Palps Sensory appendages located around the mouth.

Parasite One organism living off another and relying on it exclusively to gain its required nutrition.

Parthenogenetic Reproduced from an unfertilized egg.

Pollen Tiny grains containing the male sex cells of plants and produced by the flower's anthers.

Predator Organism that actively hunts and kills live prey.

Pronotum The hardened plate on the upper surface of an insect that covers the thorax.

Pupa Life-cycle stage between larva and adult, sometimes referred to as a chrysalis.

Rostrum Pointed, snout-like mouth structure used for feeding.

Scutellum Triangular shaped rear section of the upper thoracic plate; particularly evident in shield bugs.

Shrub Branched, woody plant.

Species Classification unit that defines organisms that are able to breed with one another and produce viable offspring.

Spinner The full adult stage of the mayfly.

Stridulation Noise produced by rubbing two parts of the body together.

Subspecies A subdivision of a species, divisions being capable of breeding with the others but often separated by geographical location.

Thorax The middle section of an insect's body.

Vagrant Appearing outside of the species' normal range, typically as a result of unusual weather patterns.

Wingspan Distance from one wing tip to the other when wings are extended.

Brimstone Moth in flight with body parts annotated

Swallowtail ■ *Papilio machaon* Wingspan 70mm

DESCRIPTION A large and iconic butterfly with wings of mainly yellow, inlaid with strong black veins and markings. The hindwings have blue and red spots and boast a pair of distinctive protruding tails, giving the butterfly its name. Larva is predominantly yellow-green with black and red markings. **SEASON** Adults are double-brooded, flying May–Jun and Aug; larva feeds Jun–Jul. **HABITAT** Lowland broads, fens and marshes. **HABITS** A busy butterfly and an energetic flier. Larva feeds on Milk-parsley. **STATUS** Widespread and common, with numbers reducing further north. In Britain, restricted to East Anglia where it is easiest to see at Strumshaw Fen and Hickling Broad.

Scarce Swallowtail ■ *Iphiclides podalirius* Wingspan 80mm

DESCRIPTION A large and strikingly marked butterfly with the unmistakable Swallowtail wing shape. The wings have distinctive black-and-white zebra-like vertical markings and subtle yellow borders. The

hindwings have blue spots and protruding tails. Larva is greenish-yellow with small dark spots. **SEASON** Adult flies May–Oct; larva feeds Jun–Oct. **HABITAT** Prefers warm, dry areas with a proliferation of Blackthorn, on which the larva feeds. **HABITS** Overwinters as a pupa. **STATUS** An extremely rare vagrant in Britain. Preferring warmer climates, Northern Europe is on the limit of its range, where it is considered common. Protected in many countries as numbers are in decline.

Large White ▪ *Pieris brassicae* Wingspan 60mm

DESCRIPTION Faithful to its name, this is a large butterfly with creamy-white upperwings and a black tip to the forewings. It is the largest of the white butterflies to be found in Britain. Sexes are distinguishable as the female has an additional 2 dark spots to each forewing. Larva is black and yellow. **SEASON** Adult flies May–Sep. Larva feeds May–Sep. **HABITAT** Frequently referred to as 'Cabbage White', it inhabits gardens, allotments, farmland and meadows. **HABITS** Larva is often found in large groups feeding on cultivated cabbages and similar brassicas, rendering them unpopular with gardeners. **STATUS** Common and widespread and one of the most familiar garden species.

Small White ▪ *Pieris rapae* Wingspan 45mm

DESCRIPTION Similar in appearance to, but appreciably smaller than, Large White (above). The upperwings are creamy-white with black tips to the forewings; underwings are yellowish. Sexes are distinguishable as females have an additional 2 dark spots to each forewing. Larva is green with small pale spots. **SEASON** Adults fly in 2 broods; Apr–May and Jul–Aug; larva feeds Jun–Sep. **HABITAT** Inhabits gardens, allotments, farmland and meadows. **HABITS** Larva is often found in large groups feeding on cultivated cabbages and similar brassicas, rendering them unpopular with gardeners. **STATUS** Common and widespread and one of the most familiar garden species.

Green-veined White ▪ *Pieris napi* Wingspan 45-50mm

DESCRIPTION Similar in size and appearance to Small White (p.17) but has obvious dark veined markings on the upperwings; greyish-green on the underwings. Sexes are distinguishable: the female is more heavily marked and bears 2 spots to the forewings compared to 1 on the male's. Larva is green with small white dots. **SEASON** Adults fly in Apr–Aug in 2 overlapping broods; larva feed May–Sep. **HABITAT** Meadows, hedgerows, roadside verges and open woodland. Males often seen congregating on mud or other nutrient-rich surfaces. **HABITS** Larva feeds on Hedge Mustard, Garlic Mustard and related plants. **STATUS** Widespread and common where suitable food plants occur.

Bath White ▪ *Pontia daplidice* Wingspan 45mm

DESCRIPTION A delicately marked, small white butterfly that displays beautiful patterning on its mainly white wings, being tipped and edged with black blotches and spots. The underside of the hindwings has a distinctive green blotchy patterning. Larva is hairy with yellow stripes running the length of the body. **SEASON** Adult flies Apr–Sep; larva feeds Jul–Aug. **HABITAT** Open disturbed habitats including wasteland and scrub. **HABITS** Larva feeds on Hedge Mustard, Sea Radish and other crucifers. **STATUS** Northern Europe marks the northern extreme of this butterfly's range where it occurs in varying numbers year on year. It is an extremely rare migrant to southern Britain.

Orange-tip ▪ *Anthocharis cardamines* Wingspan 40mm

DESCRIPTION A common, medium-sized butterfly, the males being unmistakable and give the species its name. The upperwings are rounded and white, tipped with bright orange. The female lacks the orange on the upperwing and is often confused with other white butterfly species. The undersides of the wings are mottled green and white providing efficient camouflage. Larva is green with graded white edging. **SEASON** Adult flies Apr–Jun; larva feeds May–Jul. **HABITAT** A classic spring species of open woodland, verges and rural gardens. **HABITS** Larva feeds on Cuckoo-flower. **STATUS** Common and widespread. In Britain, more common in the south.

Wood White ▪ *Leptidea sinapsis* Wingspan 40mm

DESCRIPTION A delicate-looking butterfly with rounded whitish wings, the forewings of which are dark-tipped. Larva is green with pale stripes. **SEASON** Adults are double-brooded flying May–Jun and again Jun–Jul; larva feeds Jun–Aug. **HABITAT** Meadows, open woodland and forest edges. **HABITS** Adults display a rather weak and lethargic flight. Larva feeds on the leaves of Meadow Vetchling and other members of the pea family. **STATUS** Widespread and locally common. In Britain, numbers have dwindled and only a few localized populations remain in the south and west. The almost identical **Réal's Wood White** *Leptidea reali* is a species common to southern Ireland.

Black-veined White ■ *Aporia crataegi* Wingspan 50-70mm

DESCRIPTION Very distinctive large white butterfly with strong black veins on both upper and underwings, lacking any further markings. Abdomen is black on the upperside. Larva is hairy and coloured black on the upperside, pale underneath, with orange-brown lateral stripes. **SEASON** Adult flies May–Aug; larva feeds Apr–Sep. **HABITAT** Orchards, gardens, meadows, woodland, roadsides and field edges. **HABITS** Larva feeds on Blackthorn and Hawthorn. **STATUS** Extinct in Britain since the 1920s, it is still found in mainland Europe with the exception of northern Scandinavia. Several failed attempts have been made for its re-introduction in Britain.

Brimstone ■ *Gonepteryx rhamni* Wingspan 60mm

DESCRIPTION A large butterfly with distinctive rounded wings pointed at the tips. The male is bright yellow; female colouration is more subdued. Larva is green with a pale lateral line. **SEASON** Adult hatches in Aug and hibernates, emerging again in Mar; larva feeds May–Jul. **HABITAT** Occurs in a wide range of habitats from chalk downland to woodland and is a familiar garden species. **HABITS** Commonly the first butterfly to make an appearance after the winter and can be seen on sunny days from March onwards. Larva feeds on Alder Buckthorn and Buckthorn. **STATUS** Common and widespread throughout Britain and Europe with the exception of Scotland and Scandinavia.

Clouded Yellow ▪ *Colias croceus* Wingspan 50mm

DESCRIPTION An active butterfly with yellow upperwings and dark margins. Sexes are distinguishable; male has upperwings of rich orange-yellow, pale yellow in female. Underwings are yellow and lack the dark margins. Larva is green with a lateral yellow stripe. **SEASON** Adult flies Jul–Oct; larva feeds May–Oct. **HABITAT** Can occur in most habitats but more commonly encountered in coastal regions. **HABITS** A fast flying migratory species, it arrives from southern Europe in the summer months where its range extends as far north as Scandinavia. Larva feeds on members of the pea family. **STATUS** Numbers vary from year to year, being common in some and rare in others.

Small Tortoiseshell ▪ *Aglais urticae* Wingspan 42mm

DESCRIPTION A colourful medium-sized butterfly, the upperwings are mainly reddish-orange, with black and yellow markings and a dark wing border inlaid with blue spots. The underwing is grey-brown. Larva is yellow, black and spiny. **SEASON** Adults are double- or triple-brooded and fly Mar–Oct; larva feeds May–Aug. **HABITAT** Can occur in a variety of habitats that support the larval food plant, Common Nettle. **HABITS** Adults can often be seen basking on the ground on sunny days. Adults hibernate and are sometimes found sheltering in sheds and outbuildings during the winter months. **STATUS** Common and widespread but numbers declining over recent years.

Large Tortoiseshell ■ *Nymphalis polychloros* Wingspan 70mm

DESCRIPTION Recalls Small Tortoiseshell (p.21) in appearance, but much larger in size; the blue and yellow markings on the wing margins are more subtle. Larva is black with white dots and is covered with orange spines. **SEASON** Adult flies Jun–Jul; larva feeds May–Jun. **HABITAT** Woodland, hedgerows and scrub where the main food plant is abundant. **HABITS** Larva feeds on elm and sallow. **STATUS** Widespread in suitable habitats in Europe but not usually seen in numbers. Thought to be extinct as a breeding species in Britain, now an extremely rare migrant encountered occasionally on southern coasts.

Painted Lady ■ *Vanessa cardui* Wingspan 60mm

DESCRIPTION An elegant, colourful butterfly with salmon-pink upperwings heavily patterned with black-and-white markings. Prone to fading, the colours can quickly become muted. Underwings are buffish and similarly marked to the upperwings. Larva is spiny and brown with yellow and red markings. **SEASON** Adult flies Apr–Oct; larva feeds Apr–Oct. **HABITAT** Can be found in most habitats from coastal grassland to mountain slopes. A common sight in gardens during the summer months on good migration years. **HABITS** Larva feeds primarily on thistles. **STATUS** A migrant butterfly in Britain and numbers fluctuate year on year. Widespread on the continent.

Red Admiral ▪ *Vanessa atalanta* Wingspan 60mm

DESCRIPTION A striking butterfly with black upperwings boasting a single strong red stripe and hindwing edging; the wing tips are marked with white spots. Underwings are dark marbled grey. Larva is spiny and brown with yellow markings. **SEASON** Adult flies Apr–Sep; larva feeds Apr–Nov. **HABITAT** Occurs in a range of habitats from coastal grassland to mountain slopes and a regular garden visitor. **HABITS** An active flier often seen basking on warm ground and feeding on nectar source plants. Larva feeds mainly on Common Nettle. **STATUS** Primarily a migratory species in Britain and northern Europe. Common and widespread, although numbers fluctuate annually.

Peacock ▪ *Inachis io* Wingspan 60mm

DESCRIPTION Familiar and unmistakable with red upperwings adorned with striking blue and yellow-ringed eyes to fore and hindwings. Contrastingly, the underwings are dull and appear almost black. Larva is spiny and black. **SEASON** Adult flies Mar–Sep; larva feeds May–Jul. **HABITAT** Occurs in most lowland habitats including parks and woodland and a common garden species. Adults are regularly encountered hibernating in outbuildings in the winter, emerging in the spring. **HABITS** Single brood normally hatches in July. An active flier often found basking on warm ground and feeding on nectar source plants. Larva feeds on Common Nettle. **STATUS** Common and widespread in suitable habitats.

Camberwell Beauty ■ *Aglais antiopa* Wingspan 80mm

DESCRIPTION A distinctive large butterfly with dark maroon upperwings, edged with a pale yellow border and line of bright, iridescent blue dots. Underwings are dark grey to black with the same pale border as the upperwing. Larva is spiny and black with red markings. **SEASON** Adult flies Jun–Sep; larva feeds May–Sep. **HABITAT** Lowland habitats and an occasional garden visitor. **HABITS** Most common during the months of June to September, but adults hibernate and have been encountered in most months. Larva feeds on elm, poplar and willow. **STATUS** A rare migrant to the British Isles. More common and widespread in Scandinavia and northern Europe where it breeds.

White Admiral ■ *Ladoga camilla* Wingspan 50mm

DESCRIPTION A beautiful medium-sized butterfly with black upperwings and single bold white bands on each wing. In contrast, the underside is a light chestnut colour with banding that mirrors the upperwing. Larva is green with spiny tufts of orange hairs.

SEASON Adult flies Jun–Sep; larva feeds Apr–Jun and Jul–Oct. **HABITAT** A species of deciduous woodland. **HABITS** An elegant flier, often seen gliding effortlessly along forest rides. Larva feeds on Honeysuckle and overwinters, going through a series of growth moults before pupating. **STATUS** Locally common in suitable woodland. In Britain, confined to southern regions.

Purple Emperor ■ *Apatura iris* Wingspan 65mm

DESCRIPTION A magnificent butterfly with dark brown wings and bold white band. The male has a striking purple sheen when seen from certain angles. Larva is green with distinctive 'horns' at the head end. **SEASON** Adult flies Jul–Aug; larva feeds Apr–Jun and Aug–Oct. **HABITAT** A species of deciduous woodland congregating high in the tree tops. **HABITS** At either end of the day, the males are sometimes seen feeding on animal droppings or damp earth. Larva feeds on Goat Willow and overwinters, going through a series of growth moults before pupating. **STATUS** Rare and local in suitable woodland habitats. In Britain, confined to southern regions.

Comma ■ *Polygonia c-album* Wingspan 45mm

DESCRIPTION A medium-sized butterfly with ragged edges to the wings rendering it unmistakable. The upperwings are orange-brown with darker margins and dark spots; underwings are dark grey showing a white 'comma' marking. Larva is orange-brown with tufts of spiny hairs and a bold white dorsal band. **SEASON** Adult flies Mar–Apr and Aug–Oct; larva feeds Apr–Aug. **HABITAT** Primarily a woodland species and a familiar garden visitor. **HABITS** Adults overwinter and can be encountered in any month of the year. Larva feeds on Common Nettle, Hop and elms. **STATUS** Locally common and becoming increasingly widespread with distribution extending northwards.

Queen of Spain Fritillary ▪ *Issoria lathonia* Wingspan 42mm

DESCRIPTION A medium-sized butterfly with distinctive and rather angular orange wings adorned with numerous black blotches. The underside of the forewing recalls the upperside; the underside of the hindwing is marbled brown, inlaid with characteristic silver pearl-like spots. Larva is black and spiny with red and white spots. **SEASON** Adult flies Mar–Oct; larva can occur all year in favourable conditions. **HABITAT** Favours flower-rich grasslands and meadows. **HABITS** An active and fast flier, it is highly migratory and can be encountered in a variety of habitats. Larva feeds on violets. **STATUS** Widespread and common on mainland Europe. A rare migrant species in Britain; sightings confined to the south.

Pearl-bordered Fritillary
▪ *Boloria euphrosyne* Wingspan 42mm

DESCRIPTION Adult has orange-brown upperwings inlaid with dark markings and edged with regular dark blotches. Underwing is lighter and beautifully patterned with 7 silver spots to the wing margin and 2 similar central spots. Larva is black with a pale lateral band and yellow hairy spikes. **SEASON** Adult flies May–Jun; larva feeds Jun–Oct and Mar–Apr. **HABITAT** Open deciduous woodland. **HABITS** An active, swift flying butterfly. Larva overwinters and feeds on violets. **STATUS** Widespread but only locally common with discrete colonies in suitable habitats where larval food plant proliferates.

Small Pearl-bordered Fritillary ■ *Boloria selene* Wingspan 40mm

DESCRIPTION Similar to Pearl-bordered Fritillary (p.26). Adult has orange-brown upperwings inlaid with dark markings and edged with regular dark blotches. Underwing is

lighter and beautifully patterned with seven silver spots to the wing margin and several similar spots centrally. Larva is black and hairy. **SEASON** Adult flies May–Jun; larva feeds Jun– Oct and Apr–May. **HABITAT** Open deciduous woodland and grassland. **HABITS** An active, swift-flying butterfly. Can often be seen with Pearl-bordered Fritillary but emerges a couple of weeks later in the year. Larva overwinters and feeds on violets. **STATUS** Local and common with discrete colonies in suitable habitats where larval food plant proliferates.

Dark Green Fritillary ■ *Argynnis aglaja* Wingspan 60mm

DESCRIPTION A large fritillary with rounded orange-brown upperwings and dark markings; the hindwing displays a crescent of five black spots. The butterfly gets its name from the

green underwing, inlaid with striking large white pearl- like spots. Larva is black and spiny with red spots and a faint yellow stripe along its back. **SEASON** Adult flies Jun–Aug; larva feeds Jul–Oct and Apr–May. **HABITAT** Sand dunes and chalk downland. **HABITS** An active, fast-flying butterfly often seen gliding over open ground feeding on thistles and knapweeds. Larva feeds on violets and overwinters. **STATUS** Widespread and locally common in suitable habitats throughout Europe. In Britain, favours the west.

High Brown Fritillary ■ *Fabriciana adippe* Wingspan 60mm

DESCRIPTION A large fritillary with rounded orange-brown upperwings marked with dark spots and blotches. Most easily identified by the underwings that are an orange-buff colour; the hindwings are adorned with large silvery-white markings and a crescent of chestnut spots. Larva is black and spiny with reddish spots and streaking. **SEASON** Adult flies Jun–Aug; larva feeds Mar–May. **HABITAT** Meadows, rough grassland and open, grassy woodlands. **HABITS** A fast and energetic flier that never seems to rest on warm sunny days. Larva feeds on violets. **STATUS** Widespread and locally common on mainland Europe; increasingly rare and endangered in Britain, showing a preference for the west.

Silver Washed Fritillary ■ *Argynnis paphia* Wingspan 70mm

DESCRIPTION A large fritillary with orange-brown upperwings marked with bold dark spots. As with most fritillaries, it is most easily distinguished by the underwings, the forewing being buffish and rather angular; the hindwing greenish and more rounded with distinctive silvery bands that give the butterfly its name. Larva is black and spiny with red spots and a double yellow line along its back. **SEASON** Adult flies Jun–Aug; larva feeds Aug–Nov and Mar–May. **HABITAT** Mature deciduous and coniferous woodlands. **HABITS** A powerful and busy flier. Larva feeds on Common Dog-violet. **STATUS** Widespread and locally common in suitable habitat. In Britain, confined to the south and south-west.

Marsh Fritillary

■ *Euphydryas aurinia* Wingspan 50mm

DESCRIPTION Arguably the most distinctly and heavily marked fritillary. The rather narrow wings are decorated with a series of dark lines inlaid with a mosaic of orange, yellow and brown; underwings are similar but with more subdued colours. Female is larger than male. Larva is black and spiny with tiny, star-like dots. **SEASON** Adult flies May–Jun; larva feeds Jun–Oct and Mar–Apr. **HABITAT** Grasslands, meadows, moorland and heathland. **HABITS** A rather lethargic flier that is often found at rest or basking in the sun. Larva overwinters and feeds on Devil's-bit Scabious and Plantain. **STATUS** Widespread and locally common although numbers can vary annually.

Heath Fritillary ■ *Mellicta athalia* Wingspan 45mm

DESCRIPTION A medium-sized butterfly that is a rich orange-brown with a bold lattice-work of black markings. The underwings are lighter; the forewing orange-buff with a pale

edge, the hindwing adorned with a bold central band of pale markings and also the wing edges. Larva is black with orange spines. **SEASON** Adult flies Jun–Jul; larva feeds Jul–Oct and Mar–May. **HABITAT** Woodland, heathland and rich grassland. **HABITS** A sluggish flier that can often be found basking on shrubs. Larva feeds on Common Cow-wheat and Germander Speedwell. **STATUS** Widespread and common on mainland Europe. In Britain, confined to a limited number of sites in the extreme south.

Glanville Fritillary ■ *Melitaea cinxia* Wingspan 40mm

DESCRIPTION A small fritillary with orange-brown upperwings with bold black markings and 4–5 small black dots on the hindwing. The underwings are lighter with distinctive

concentric bands of orange and creamy white to the hindwing. Larva is black and bristly with a red head. **SEASON** Adult flies May–Jun; larva feeds Mar–Apr and Jul–Oct. **HABITAT** Grassy areas such as coastal undercliffs, verges and hill slopes. **HABITS** A sun-loving butterfly only active in bright sunshine. Occasional second brood on the wing during Aug. Larva feeds on Ribwort Plantain and overwinters. **STATUS** Widespread and fairly common. In Britain, restricted to the Isle of Wight and the south coast.

Duke of Burgundy ■ *Hamearis lucina* Wingspan 25mm

DESCRIPTION A small butterfly that, although unrelated, shows similar markings to a fritillary, the upperwings beautifully patterned with orange and brown and subtle light

edging. The underwings are lighter, the forewing marked with dark spots; the hindwing with a series of large, pearl-like white spots. Larva is brown and slightly hairy. **SEASON** Adult flies May–Jun; larva feeds Jun–Jul. **HABITAT** Meadows and woodland margins where larval foodplant occurs. **HABITS** Flight is fast and busy. Occasional second brood in August. Larva feeds on Primrose and Cowslip and is nocturnal. **STATUS** Widespread and locally common in Europe. In Britain, restricted to England.

Speckled Wood ▪ *Pararga aegeria* Wingspan 45mm

DESCRIPTION A common, medium-sized butterfly, rich brown in colour with a busy pattern of bold yellow-buff spots; the underwings are similarly marked. To the south of its

range, a second colour form can occur being orange and brown with 3 distinctive eyespots; 1 on the forewing, 2 on the hindwing resembling a Wall Brown (below). Larva is green. **SEASON** Adult flies Mar–Oct in 2 broods; larva feeds May–Oct and Mar. **HABITAT** Woodland, hedgerows and an occasional garden visitor. **HABITS** A familiar butterfly that is often on the wing in overcast as well as sunny conditions. Larva feeds on various woodland grasses and overwinters. **STATUS** Widespread and locally common.

Wall Brown ▪ *Lasiommata megera* Wingspan 45mm

DESCRIPTION A medium-sized butterfly with orange-brown upperwings that recalls a fritillary in pattern. The single eyespot on the forewing and a series of similar spots to the hindwings are the distinctive features of this species. The underwings are lighter; orange-buff on the forewing, grey-brown on the hindwing and display the eyespots of the upperwings. Larva is green with light stripes. **SEASON** Adult flies May–Sep in 2 broods; larva feeds Mar–Apr and Jun–Oct. **HABITAT** Dry grassland, heathland, cliffs and hillsides. **HABITS** A sun-loving butterfly often found basking on rocks and bare ground. Larva feeds on grasses and overwinters. **STATUS** Widespread but numbers in decline.

Scotch Argus ■ *Erebia aethiops* Wingspan 40mm

DESCRIPTION A medium-sized butterfly with broad wings. The upperwings are a rich dark brown, decorated with orange bands towards the edges and inlaid with distinctive

eyespots with white highlights. The underside of the forewing recalls the upperside, the hindwing is dull brown with a greyish submarginal band. Larva is buffish with thin dark stripes. **SEASON** Adult flies Jul–Sep; larva feeds Mar–Jun and Aug–Oct. **HABITAT** A species of cool climates, preferring mountain grassland and moorland. **HABITS** A slow-flying butterfly that is often encountered basking on vegetation. Larva feeds on Purple Moor-grass. **STATUS** Locally common in upland areas of Europe. In Britain, restricted to Scotland and Cumbria.

Mountain Ringlet ■ *Erebia epiphron* Wingspan 32mm

DESCRIPTION A small, delicate butterfly with brown wings that can vary in shade with local populations. The upperwings are marked with an orange band inlaid with eyespots.

Underwings are brown, the forewing yellowish with eyespots, the hindwing brown with a greyish submarginal band. Larva is buffish with a dark stripe along its back. **SEASON** Adult flies Jun–Jul; larva feeds Apr–May and Aug–Sep. **HABITAT** Mountains and upland moors. **HABITS** Flies only in direct sunshine, immediately going to ground in cloud cover. Larva feeds on Mat-grass and overwinters. **STATUS** Widespread and locally common in suitable habitats. In Britain, restricted to central Scotland and Cumbria.

Marbled White ■ *Melanargia galathea* Wingspan 55mm

DESCRIPTION An unmistakable, largish butterfly that has distinctively patterned upperwings of bold black veins and black-and-white blotches. The underwings are similar, but with subdued markings and a yellowish tinge. The larva is light green or brown with a dark dorsal line. **SEASON** Adult flies Jun–Sep; larva feeds Apr–Jun. **HABITAT** Undisturbed grassland including downland, road verges, field margins and open woodland. **HABITS** Adult has a rather sluggish flight and is often seen feeding in the open on thistles and knapweeds. Larva feeds on grasses, is strictly nocturnal and overwinters. **STATUS** Widespread and locally common. In Britain, restricted to southern and central areas.

Grayling ■ *Hipparchia semele* Wingspan 50mm

DESCRIPTION A butterfly that rarely shows its upperwings, preferring to hold its wings in a closed position and angled to produce minimal shadow. Upperwings, when revealed, are buffish-brown with a yellowish submarginal band marked with eyespots. The underside of the forewing has an orange patch with 2 eyespots, the hind underwing is marbled grey-brown affording excellent camouflage. Regional variation occurs. Larva is light green to brown with dark stripes. **SEASON** Adult flies Jul–Sep; larva feeds Mar–Jun and Aug–Sep. **HABITAT** Heathland, sand dunes and coastal grassland. **HABITS** An active, sun-loving species difficult to spot at rest. Larva feeds on grasses and overwinters. **STATUS** Widespread and locally common.

Ringlet ▪ *Aphantopus hyperantus* Wingspan 48mm

DESCRIPTION A largish butterfly with distinctive dark brown upperwings. Eyespots are present on both upper and underwings, but can be hard to spot on the darker upperwings. Underwings are a lighter brown and the contrasting dark eyespots are ringed yellow with central white highlights. Local variation in the number, size and shape of the eyespots can occur. Larva is hairy, light green or brown with a dark dorsal stripe. **SEASON** Adult flies Jun–Jul; larva feeds Apr–Jun and Jul–Sep. **HABITAT** Open woodland, roadside verges, meadows, hedgerows and an occasional garden visitor. **HABITS** Larva feeds on grasses, is strictly nocturnal and overwinters. **STATUS** Widespread and common.

Gatekeeper ▪ *Pyronia tithonus* Wingspan 40mm

DESCRIPTION Has warm brown upperwings inlaid with orange; underwings are similar but lighter in colour. A distinctive dark eyespot on each forewing, visible on both the upperwing and underwing, is thought to deflect attention from predators. Sexes are distinguishable, male is smaller and bears additional dark, scent-producing patches to each upperwing. Larva is pale brown. **SEASON** Adult flies Jul–Aug; larva feeds Mar–Jun and Aug–Sep **HABITAT** Waysides, hedgerows, field margins and grassland scrub. **HABITS** An active butterfly often found feeding on bramble flowers. Larva feeds on meadow grasses and overwinters. **STATUS** Widespread and locally common. Its range is thought to be extending northwards.

Meadow Brown ■ *Maniola jurtina* Wingspan 50mm

DESCRIPTION A common butterfly with brown upperwings with a patch of orange on the forewing containing one eyespot. The size of the orange patch allows gender separation, the female having a larger patch than the male. The underside of the forewings are orange and buff with an eyespot; the hindwing buff with a wide, greyish marginal band and small dark spots. Larva is green. **SEASON** Adult flies Jun–Sep; larva feeds Apr–Jun and Jul–Sep. **HABITAT** One of the most common species, it is encountered in grassland, meadows, field margins, woodland rides, road verges and an occasional garden visitor. **HABITS** Larva feeds on grasses nocturnally and overwinters. **STATUS** Common and widespread.

Large Heath ■ *Coenonympha tullia* Wingspan 40mm

DESCRIPTION A smallish butterfly that rarely reveals its upperwings, typically found at rest with wings upright. When glimpsed, upperwing is orange-buff with variable subtle eyespots. Underwings are orange-brown on the forewing, greyish on the hindwing, with eyespots that vary in number and intensity. Larva is green with lighter, yellowish stripes. **SEASON** Adult flies Jun–Jul; larva feeds Apr–May and Jul–Sep. **HABITAT** A hardy upland butterfly of acid bogs and wet moorland. **HABITS** Only flies in bright sunshine. Larva feeds on White Beak-sedge and overwinters for up to 2 seasons. **STATUS** Widespread but local in upland areas. In Britain found only from central Wales northwards.

Small Heath ■ *Coenonympha pamphilus* Wingspan 30mm

DESCRIPTION A small butterfly that rarely shows its upperwings, typically found at rest with wings upright. When revealed, upperwings are orange-brown in colour. More diagnostic underwings are orange with a single dark eyespot to the forewing, marbled grey, brown and buff to the hindwing. Larva is green with light striping. **SEASON** Adult flies May–Sep in 2 broods; larva feeds Mar–Apr and Jun–Sep. **HABITAT** Grassland, heathland, meadows and sand dunes. **HABITS** A charming little butterfly with a fluttering flight. Larva feeds on grasses nocturnally and overwinters. **STATUS** Widespread and locally common, although numbers thought to be in decline. In Britain, common only in the south.

Purple Hairstreak ■ *Quercusia quercus* Wingspan 40mm

DESCRIPTION A delightful butterfly that will show its upperwings on occasion. A deep sooty-brown colour, the male has a purple sheen that can be striking when seen in certain light. Underwings are also distinctively marked being buffish-grey with a thin, white

submarginal band and orange wing edgings. Larva is reddish-brown. **SEASON** Adult flies Jun–Aug; larva feeds Mar and Jul–Sep. **HABITAT** Mature oak woodland. **HABITS** A common species, it spends the majority of its time in the tree canopy. Can be seen at ground level on dull days, especially after rain. Larva feeds on oak nocturnally and overwinters. **STATUS** Widespread and locally common.

White-letter Hairstreak ▪ *Strymondia w-album* Wingspan 35mm

DESCRIPTION A small butterfly that seldom reveals its uniform dark brown upperwings. The underwings are a rich brown with a diagnostic thin, W-shaped white band, giving

the butterfly its name, and a crescent-shaped orange band to the hindwing edge. The hindwing displays an obvious tail-streamer. Larva is green and slug-like. **SEASON** Adult flies Jun–Aug; larva feeds Mar–Apr and Jul–Sep. **HABITAT** Deciduous woodland and mature hedgerows that contain the larval foodplant. **HABITS** An elusive, busy butterfly that mostly flies around in treetops, occasionally comes to bramble flowers. Larva feeds on elms and overwinters. **STATUS** Widespread but local owing to Dutch Elm disease.

Black Hairstreak ▪ *Strymondia pruni* Wingspan 35mm

DESCRIPTION A small butterfly that rarely shows its rich brown upperwings when at rest. The underwings bear an obvious tail streamer and are also a rich brown. A thin white line and a strong, orange submarginal band, decorated with white-edged black spots, adorn the hindwing. Larva is green and slug-like. **SEASON** Adult flies Jun–Jul; larva feeds Mar–Jun. **HABITAT** Mature hedgerows, woodland margins and scrub. **HABITS** A sluggish flier that lives in colonies and is vulnerable to habitat loss. Can sometimes be found crawling over hedgerow vegetation. Larva feeds on Blackthorn. **STATUS** Widespread but extremely local. In Britain, restricted to a handful of sites in the East Midlands.

Brown Hairstreak ■ *Thecla betulae* Wingspan 40-50mm

DESCRIPTION A medium-sized butterfly with dark brown upperwings. The males display a distinct orange patch on the forewing making the sexes distinguishable. The underwings are orange-brown and marked with a fine, black-edged white line; hindwings display a delicate tail streamer. Larva is bright green and slug-like. **SEASON** Adult flies Jul–Sep; larva feeds May–Jun. **HABITAT** Mature hedgerows, woodland edges and scrub where larval foodplant is abundant. **HABITS** A rather lazy, slow-flying butterfly that is often found walking over foliage. Forms discrete colonies. Larva feeds on Blackthorn. **STATUS** Widespread but extremely local. In Britain, restricted to southern England and Wales.

Green Hairstreak ■ *Callophrys rubi* Wingspan 40mm

DESCRIPTION A small butterfly that rarely shows its rich brown coloured upperwings when at rest. Underwings are a vibrant green with a broken white line to both forewing and hindwing. Its colouration is a very effective camouflage making it difficult to spot when not in flight. Larva is green with light stripes and slug-like. **SEASON** Adult flies Apr–Jun; larva feeds May–Jul. **HABITAT** Occurs in a variety of scrubby habitats including downland, heathland, hedgerows, woodland margins and rough grassland. **HABITS** A busy flier that lives in colonies. Larva feeds on various low-growing shrubs such as Bird's-foot Trefoil, Bilberry and Broom. **STATUS** Widespread and locally common.

Small Copper ■ *Lycaena phlaeas* Wingspan 40mm

DESCRIPTION A small, strikingly beautiful butterfly. The upperside of the forewing is bright orange with brown edging and inlaid with brown spots. Contrastingly, the hindwing is brown with an orange submarginal band. The underwings are similarly marked, the forewing being orange with black spots and buff wing edge, the hindwing a uniform buff. Some local colour variation can occur. Larva is green and slug-like. SEASON Adult flies Apr–Oct in 2 broods; larva feeds Apr–Sep . HABITAT Grassland, wasteland, heathland, woodland margins and roadside verges. HABITS A familiar fast flyer that forms discrete colonies. Larva feeds on sorrels, some overwinter. STATUS Extremely widespread and locally common.

Common Blue ■ *Polyommatus icarus* Wingspan 32mm

DESCRIPTION A small butterfly. The male has bright blue upperwings tinged with violet; the female is rich brown with orange submarginal spots to fore and hindwings.

The underside is similar in both sexes, being grey-brown with black and orange white-ringed spots. Larva is green and slug-like. SEASON Adult flies May–Sep in 2–3 successive broods; larva feeds Apr and Jun–Sep. HABITAT Meadows, grassland, roadside verges, sand dunes and waste ground where larval foodplant is common. HABITS A familiar butterfly that is at its most active in bright sunshine. Larva feeds on trefoils and clovers, some overwinter. STATUS Common and widespread.

Chalkhill Blue ■ *Polyommatus coridon* Wingspan 40mm

DESCRIPTION The male is uniquely pale blue to the upperwings, with a dark brown margin and white chequered fringe that is also a feature of the female. The female is dark brown with orange and black eyespots to the hindwings. Both sexes have grey-brown underwings with dark spots. Larva is green with yellow stripes. **SEASON** Adult flies Jul–Aug; larva feeds Apr–Jun. **HABITAT** Chalk and limestone grassland. **HABITS** Most active in bright sunshine when males can occur in abundance, flying close to the ground searching for a mate. Larva feeds on Horseshoe Vetch. **STATUS** Widespread and locally common. In Britain its range is restricted to the south.

Adonis Blue ■ *Polyommatus bellargus* Wingspan 32mm

DESCRIPTION A classic chalk downland butterfly, the male of which is a brilliant iridescent blue with black-and-white chequered wing margins. The female has brown upperwings with orange spots and a variable blue tinge. Underwings of both sexes are grey-brown with black spots. Larva is green with yellow stripes.

SEASON Adult flies May–Sep in successive broods; larva feeds Mar–Apr and Jun–Sep. **HABITAT** Occurs only on chalk downland. **HABITS** A lover of warm, sheltered spots flying low over vegetation seeking the inconspicuous females. Larva feeds on Horseshoe Vetch, some overwinter. **STATUS** Widespread and locally common. In Britain, restricted to the south. Vulnerable to habitat loss.

Holly Blue ■ *Celastrina argiolus* Wingspan 30mm

DESCRIPTION A small, familiar blue butterfly, the sexes similarly marked with blue upperwings with variable dark wing margins that are generally much broader in the female. The underwings are lighter being a uniform blueish-white, decorated with subtle black spots. Larva is green and slug-like. **SEASON** Adult flies in 2 broods Apr–May and Aug–Sep; larva feeds May–Sep. **HABITAT** Occurs in a variety of habitats where larval foodplants are in abundance, including woodland and parkland. **HABITS** A familiar garden visitor. Larva feeds on holly and ivy. **STATUS** Common and widespread. In Britain, more common in the south. Numbers fluctuate annually owing to larval parasitism.

Small Blue ■ *Cupido minimus* Wingspan 25mm

DESCRIPTION A tiny butterfly that, in contrast to its name, has sooty brown upperwings. Only the male displays a tinge of blue near the wing base, making the sexes distinguishable.

The underwings in both sexes are pale grey adorned with small black spots. Larva is green with a small black head. **SEASON** Adult flies May–Aug in up to 2 broods; larva feeds Jun–Sep. **HABITAT** Chalk and limestone downland, rich grassland, road embankments and open woodland with an abundance of the larval foodplant. **HABITS** A busy flying butterfly active only during sunny periods. Larva feeds on Kidney Vetch and overwinters. **STATUS** Widespread and locally common.

Silver-studded Blue ■ *Plebejus argus* Wingspan 25-30mm

DESCRIPTION A small butterfly, the male has blue upperwings with white margins and a dark submarginal band; the female has brown upperwings with orange submarginal spots.

Underwings in both sexes are pale grey-blue, the scales reflecting light and giving the butterfly its name. A submarginal band of black and orange spots edges each underwing. Larva is green with a dark dorsal stripe. **SEASON** Adult flies May–Sep in up to 2 broods; larva feeds Mar–May. **HABITAT** Managed heathland and calcareous grassland. **HABITS** A busy, fast-flying butterfly that lives in tight colonies. Larva feeds on heather and gorse. **STATUS** Widespread but common only locally. In Britain, restricted to southern England and Wales.

Large Blue ■ *Maculinea arion* Wingspan 40mm

DESCRIPTION One of the largest blue species, both sexes have blue upperwings with dark margins and dark spots; larger and more pronounced in the female. Underwings are pale grey-brown with a blue tinge to the base of hindwing and marked with dark spots and chequered margins. Larva is green and slug-like. **SEASON** Adult flies Jul; larva feeds Apr–May and Jul–Sep. **HABITAT** Grazed grassland with an abundance of the larval foodplant. **HABITS** Larva feeds on Thyme and in the latter stages of development parasitizes Red Ants in a 'cuckoo-like' manner. **STATUS** Widespread but scarce and declining. In Britain, restricted to a few sites in the south-west.

Brown Argus ■ *Aricia agestis* Wingspan 25mm

DESCRIPTION A small butterfly, both sexes of which have brown upperwings that display a subtle blue iridescence in certain light. Bold orange spots form submarginal bands to both fore and hindwings. The underwings are grey-brown with white-ringed black spots, and orange submarginal spots. Larva is green with a yellow dorsal stripe. **SEASON** Adult flies May–Aug in 2 broods; larva feeds Jun–Sep and Apr. **HABITAT** Calcareous downland, heathland and open woodland. **HABITS** A lover of the warmth and often found on sheltered, south-facing slopes. Larva feeds on Common Rock-rose and Common Stork's-bill and overwinters. **STATUS** Widespread and locally common. In Britain, restricted to the south.

Northern Brown Argus ■ *Aricia artaxerxes* Wingspan 25mm

DESCRIPTION Recalls Brown Argus (above). Upperwings are rich brown with bright orange submarginal spots that are fainter than Brown Argus. Those that occur in Britain bear a distinctive white spot on each forewing. Underwings are grey-brown with white-ringed black spots and orange submarginal spots. Larva is green and slug-like. **SEASON** Adult flies Jun–Aug; larva feeds Apr–May and Jul–Sep. **HABITAT** Calcareous grassland and areas of limestone with an abundance of the larval foodplant. **HABITS** Larva feeds on Common Rock-rose and overwinters. **STATUS** Widespread but with localized populations in upland areas or northern latitudes. In Britain, restricted to northern England and Scotland.

Small Skipper ■ *Thymelicus sylvestris* Wingspan 25mm

DESCRIPTION A small butterfly with warm orange-brown upperwings and dark narrow wing margins; underwings are orange-buff. Sexes can be distinguished as the male possesses a sex brand in the form of a thin dark line on the forewings. Easily confused with the Essex Skipper (below) but distinguished from it by the brown underside to antennal tip. Larva is light green. **SEASON** Adult flies May–Sep; larva feeds Apr–Jun and Aug–Sep. **HABITAT** Meadows, rough grassland and roadside verges. **HABITS** An active butterfly often found basking on vegetation or making short, busy flights in tall grass. Larva feeds on meadow grasses nocturnally and overwinters. **STATUS** Widespread and common. Absent from Scotland and Ireland.

Essex Skipper ■ *Thymelicus lineola* Wingspan 25mm

DESCRIPTION Recalls the Small Skipper (above) but distinguished from it by the black underside to antennal tip. Adult has warm orange-brown upperwings with narrow dark wing margins; underwings are orange-buff. Sexes can be distinguished; the male possesses a sex brand in the form of a thin dark line on the forewings. Larva is light green. **SEASON** Adult flies May–Aug; larva feeds Mar–Apr and Jul–Sep. **HABITAT** Meadows, coastal grassland and roadside verges. **HABITS** An active butterfly with a busy, buzzing flight. Can often be found visiting thistles and clovers. Larva feeds on meadow grasses nocturnally and overwinters. **STATUS** Widespread and relatively common. In Britain, restricted to the south.

Lulworth Skipper ■ *Thymelicus acteon* Wingspan 28mm

DESCRIPTION A small, well-marked butterfly with olive-brown upperwings. Forewing has a distinctive 'paw-print'-shaped crescent of pale yellow spots, brighter on the female than the male. The male has a sex brand on the forewing in the form of a thin dark line. Larva is light green. **SEASON** Adult flies May–Jul; larva feeds Apr–May and Aug–Sep. **HABITAT** Meadows and rough grassland. **HABITS** An active butterfly with a rapid, buzzing flight. A nectar lover, often found flitting between the flowers of thistles. Larva feeds on meadow grasses nocturnally and overwinters. **STATUS** Widespread and locally common, less frequent in the north. In Britain, restricted to Dorset and east Devon.

Large Skipper ■ *Ochlodes sylvanus* Wingspan 34mm

DESCRIPTION One of the largest of the golden skippers. Upperwings are a warm orange-brown with broad dark wing margins and faint pale spots. Male has a sex brand on the forewing in the form of a thin dark line. Underwings are yellow-brown. Larva is green and maggot-like. **SEASON** Adult flies May–Sep; larva feeds Apr–May and Jul–Sep. **HABITAT** Grasslands, meadows, open woodland, field margins and scrub. Seeks sheltered locations. **HABITS** Often found basking holding its wings at an acute angle. Larva feeds on meadow grasses nocturnally and overwinters. **STATUS** Widespread and common. Absent from Scotland and Ireland.

Silver-spotted Skipper ■ *Hesperia comma* Wingspan 34mm

DESCRIPTION An iconic chalk downland butterfly similar in appearance to the Large Skipper (p.45). Upperwings are dark brown and distinctively marked with pale spots and broad, dark wing margins. Male has a sex brand on the forewing in the form of a thin dark line. Underwing is greenish-brown and adorned with silver-white spots giving the species its name. Larva is green with a black head. **SEASON** Adult flies Jul–Aug; larva feeds Mar–Jul. **HABITAT** Restricted to chalk downland. **HABITS** A sun-loving, fast-flying butterfly that can often be found basking or feeding on thistles. Larva feeds on Sheep's-fescue. **STATUS** Widespread and extremely local. In Britain, restricted to the south.

Dingy Skipper ■ *Erynnis tages* Wingspan 25mm

DESCRIPTION A small moth-like butterfly that when freshly emerged has upperwings subtly marked with a brown and grey patterning, quickly lost with wear and turning a

uniform dark grey-brown. Underwings are reddish-brown. Larva is green with a black head. **SEASON** Adult flies May–Aug; larva feeds Apr and Jun–Sep. **HABITAT** Grassland, chalk downland and open woodland. A fast-flying butterfly that stays close to the ground. **HABITS** A sun-lover that is often found basking on bare ground. Larva feeds nocturnally on Bird's-foot Trefoil and related plants and overwinters. **STATUS** Widespread and locally common. Numbers declining due to habitat loss.

Grizzled Skipper ■ *Pyrgus malvae* Wingspan 20mm

DESCRIPTION A tiny butterfly with beautifully marked upperwings of dark grey-brown, bold white spots and a chequered wing border. Underwings are reddish-brown with large blotches of white. Larva is green with a black head. **SEASON** Adult flies Apr–Jun; larva feeds Jun–Jul. **HABITAT** Undisturbed grassland and woodland clearings. **HABITS** An active and fast-flying butterfly, it is also a lover of the sun and often found basking on bare ground. To the south of its range, a second brood can occur, flying Jul–Aug. Larva feeds on Wild Strawberry and cinquefoils. **STATUS** Widespread and locally common. Absent from Scotland and Ireland.

Chequered Skipper ■ *Carterocephalus palaemon* Wingspan 25mm

DESCRIPTION A strikingly marked little butterfly. The boldly marked upperwings of dark brown and buffish-orange spots create a chequerboard effect, giving the butterfly its name.

The underwings are orange-buff with a number of large lighter coloured spots. Larva is green and slender. **SEASON** Adult flies May–Jul; larva feeds Jun–Sep. **HABITAT** Grassy scrub and open woodland. **HABITS** An active and fast flyer on sunny days, taking to cover when overcast. A sun-loving species often found basking. Larva feeds on Purple Moor-grass and overwinters. **STATUS** Widespread and locally common. Absent from Britain with the exception of a few open birch woods in north-west Scotland.

Ghost Moth ■ *Hepialus humuli* Wingspan 45-50mm

DESCRIPTION A largish distinctively pale moth, the male of which has white wings with subtle dark veins. The female is yellow with distinctive orange markings. At rest the wings are held in a tent-like pose. Larva is whitish and maggot-like. **SEASON** Adult flies Jun–Aug. **HABITAT** Meadows, downland, field margins and gardens. **HABITS** Flies from dusk onwards and comes to light. Large concentrations of males are sometimes seen hovering in a ghost-like manner close to vegetation, giving the species its name. After mating, females disperse their eggs in flight. Larva feeds underground on the roots of grasses and small plants and overwinters. **STATUS** Widespread and common.

Six-spot Burnet ■ *Zygaena filipendulae* Wingspan 25-40mm

DESCRIPTION A beautifully marked moth with dark iridescent greenish-blue wings, each distinctively marked with 3 pairs of red spots allowing separation from other species of burnet. The hindwings are the same shade of red as the forewing spots and are revealed in flight. Colouration acts as a warning to predators of its toxic nature. Larva is yellow with black spots. **SEASON** Adult flies Jun–Aug. **HABITAT** Downland, meadows, woodland rides, clifftops and other flower-rich habitats. **HABITS** A day-flying species associated with hot and sunny summer days. Larva feeds on Bird's-foot Trefoil and overwinters. **STATUS** Widespread and common.

Hornet Moth

■ *Sesia apiformis* Wingspan 35-45mm

DESCRIPTION As its name suggests, this species recalls a true Hornet (p.138) having a plump yellow and black striped abdomen, transparent wings and thickened antennae. Larva is pale and grub-like. **SEASON** Adult flies Jul–Aug. **HABITAT** River valleys, marshes and similar areas; linked with the presence of mature Black Poplar. **HABITS** A rarely seen, day-flying species associated with warm and sunny days. Larva feeds exclusively on the roots and trunk of Black Poplar, burying into the wood where it feeds for at least 2 years before emerging as an adult. **STATUS** Widespread and locally distributed; more common in the south.

Bee Moth ■ *Aphomia sociella* Wingspan 20-44mm

DESCRIPTION The male of this species has a rather two-toned appearance, being pinkish at the head, grading to drab brown towards the tips of its wings. The female is uniformly drab brown with dark spots. Wings appear 'rolled' rather than folded when at rest. Larva is white and grub-like. **SEASON** Adult flies Jun–Aug. **HABITAT** Most common in woodland, gardens and scrub where nesting bees and wasps occur. **HABITS** A nocturnal flier, the adults can be attracted to light. Eggs are laid in bee and wasp nests, the hatching larva feeding and pupating within the honeycomb cells. **STATUS** Widespread and common.

Green Oak Tortrix ▪ *Tortrix viridana* Wingspan 18-22mm

DESCRIPTION A relatively easy moth to identify owing to its uniformly bright green forewings; hindwings are greyish. Folds its wings into a shield shape when at rest. Abdomen and limbs are reddish-brown. Larva is green and grub-like with a dark head. **SEASON** Adult flies Jun–Aug. **HABITAT** Mature deciduous woodland. **HABITS** Adult flies at night and can be attracted to light. A Micro-moth and a member of the Leaf-rollers family owing to the habit of the larva rolling a leaf into a tube around itself for protection. Larva feeds on the leaves of oak trees. **STATUS** Widespread and common.

Small Magpie ▪ *Eurrhypara hortulata* Wingspan 25-28mm

DESCRIPTION An attractively marked, and easily recognized moth. Upperwings are boldly marked, being white with large black spots and broad black wing margins. The body is a bold yellow and black with a yellow flush to the wing junction. Larva is greenish-yellow and grub-like. **SEASON** Adult flies May–Aug. **HABITAT** Woodland, commons, gardens, wasteland and scrub. **HABITS** A night-flying Micro-moth that can be attracted to light. Adults are sometimes disturbed from vegetation during the day. Larva feeds on nettles, rolling a leaf into a tube shape around itself. **STATUS** Common and widespread. In Britain, common only to southern and central regions.

Mother of Pearl ▪ *Pleuroptya ruralis* Wingspan 25-38mm

DESCRIPTION One of the largest of the Micro-moths. At rest this species holds its wings out flat with all four on display. Upperwings are buffish with a busy pattern of lines and veins; a pinkish pearl-like sheen recalling the inside of an oyster shell gives the moth its name. Larva is green and grub-like. **SEASON** Adult flies Jun–Aug. **HABITAT** Wasteland, woodland, gardens and other locations where nettles occur. **HABITS** Adults fly at night and are attracted to light. Can sometimes be found resting on vegetation during the day and will fly if disturbed. Larva feeds on nettles. **STATUS** Widespread and common.

White Plume Moth ▪ *Pterophorus pentadactyla* Wingspan 26-34mm

DESCRIPTION One of the largest and most distinctive of all the plume moths. As its name suggests, this species is entirely white; the forewings are deeply divided into 2 finely feathered plumes, the hindwings into a further 3. At rest it holds its wings open and flat. Larva is light green with yellow spots and spiny. **SEASON** Adult flies Jun–Aug. **HABITAT** Dry grassland, wasteland and gardens. **HABITS** Adults are most often found flying at dusk and dawn and are attracted to light. Sometimes a second generation in September can occur. Larva feeds on bindweed and overwinters. **STATUS** Widespread and fairly common.

Oak Eggar ▪ *Lasiocampa quercus* Wingspan 45-70mm

DESCRIPTION A broad-winged moth that can vary considerably in colour from a light buffish-yellow to dark brown. Most forms display a small, dark-ringed white spot and broad, light band on each forewing. The female is darker in colour and much larger than the male. Larva is brown and hairy. **SEASON** Adult flies May–Aug. **HABITAT** Deciduous woodland, hedgerows, moorland and commons. **HABITS** Females fly at night and are attracted to light: males fly with a zig-zagging flight during sunny days attracted to the female's scent. Larva feeds mainly on heather and can feed for 1 or 2 seasons. **STATUS** Widespread and locally common.

The Drinker
▪ *Euthrix potatoria* Wingspan 30-35mm

DESCRIPTION An attractive moth with broad, rounded wings and a distinctive 'snout'. Sexes are distinguishable. The male is orange-brown with 2 white spots and a dark diagonal stripe to the forewings; the female is larger and similarly marked but with lighter yellow-buff wings. Larva is hairy, large and brown with yellow stripes. **SEASON** Adult flies Jul–Aug. **HABITAT** Occurs in a number of grassy habitats including commons, open woodland, moorland and fens. **HABITS** Flies at night and is attracted to light. Larva feeds on grasses and is said to drink dewdrops, giving the species its name. **STATUS** Widespread and common. In Britain, more common in the south.

Lappet ■ *Gastropacha quercifolia* Wingspan 40mm

DESCRIPTION A spectacular moth that mimics autumn leaves, holding its wings in a tent-like manner when at rest. The adult possesses broad, rich reddish-brown wings that have subtle wavy dark bands and scalloped margins. The palps on the moth's head project as a distinctive 'snout'. The larva is among the largest being up to 10cm long, its brown and white markings providing perfect camouflage. **SEASON** Adult flies Jun–Aug. **HABITAT** Hedgerows, scrub, gardens, downland and open woodland. **HABITS** Flies at night and is attracted to light. Larva feeds on Hawthorn and Blackthorn. **STATUS** Widespread and locally common. In Britain, common only in the south.

Emperor Moth ■ *Saturnia pavonia* Wingspan 40-60mm

DESCRIPTION A beautiful and unmistakable moth, the sexes of which differ. The male has brown forewings and orange hindwings, each patterned with dark wavy lines and a large conspicuous eyespot. The female has grey fore and hindwings that are similarly marked with larger eyespots. Larva is green with black rings. **SEASON** Adult flies May–Aug. **HABITAT** Found in a variety of habitats including heathland, moorland and open woodland where larval food plants flourish. **HABITS** The male is a day-flying moth, the female nocturnal and can be attracted to light. Larva feeds on heathers, Bramble, Hawthorn, sallow and related plants. **STATUS** Widespread and locally common.

Pebble Hook-tip ▪ *Drepana falcataria* Wingspan 28mm

DESCRIPTION A well marked moth that has a distinctive hooked tip to the forewings, giving the species its name. Wing colour is variable, ranging from orange-brown to grey-buff. The forewings have a prominent dark line, several faint wavy markings and a small 'pebble-like' eyespot. Wings are open at rest and resemble a leaf. Larva is mainly green and spiny. **SEASON** Adult flies May–Aug in 2 broods. **HABITAT** Woodland, heathland, commons and gardens where Silver Birch proliferates. **HABITS** Adult flies at night and is attracted to light. Larva feeds on birch. **STATUS** Widespread and locally common. In Britain, more common in central and southern areas.

Chinese Character

▪ *Cilix glaucata* Wingspan 12-13mm

DESCRIPTION A small and interestingly marked moth that has white wings with blotched markings of brown, grey and purplish-blue. Folds its wings into a tent-like posture when at rest and looks remarkably like a bird dropping, affording very effective camouflage. Larva is brownish and wedge-shaped. **SEASON** Adult flies May–Aug in 2 broods. **HABITAT** Open woodland, hedgerows, gardens, wasteland and common land where larval food plants occur. **HABITS** Adult flies at night and is attracted to light. Larva feeds on Hawthorn, Blackthorn and Bramble. **STATUS** Widespread and locally common. In Britain, more common in central and southern areas.

Peach Blossom ▪ *Thyatira batis* Wingspan 17mm

DESCRIPTION A strikingly marked moth that is most unlike any other species having brown wings and an unusual pattern of large pink, buff and white petal-like spots. Larva is reddish-brown with a series of triangular 'humps' along its back. **SEASON** Adult flies May–Aug sometimes in 2 broods. **HABITAT** Woodland, common land, gardens and wasteland where larval foodplant proliferates. **HABITS** Adult flies at night and is attracted to light; hides in ground cover by day. Larva feeds on Bramble mainly at night. Overwinters as a pupa. **STATUS** Widespread and locally common. In Britain, common only in central and southern regions.

March Moth ▪ *Alsophila aescularia* Wingspan 19mm

DESCRIPTION A narrow-winged moth, the male's upperwings have a subtle pattern of grey and brown, with a lighter coloured, jagged line across the centre of the forewings. At rest, the wings are folded tightly with one forewing overlapping the other. The female is wingless. Larva is green. **SEASON** Adult flies Mar–Apr. **HABITAT** Woodland, mature hedgerows, common land, orchards and gardens. **HABITS** Adult flies at night and is attracted to light. The female crawls around on tree trunks at night. Larva feeds on oak, birch, Hawthorn and other deciduous trees. **STATUS** Widespread and common. In Britain, more common in central and southern regions.

Large Emerald ■ *Geometra papilionaria* Wingspan 40-50mm

DESCRIPTION A large and very colourful moth which, as its name suggests, is a rich emerald green in colour with a pair of lighter coloured, subtle scalloped lines across fore and hindwings. Wings are held flat and outstretched at rest. Colour is prone to swift fading due to wear. Larva is large and brown or green with brownish raised spots. **SEASON** Adult flies Jun–Aug. **HABITAT** Open woodland, heathland and common land. **HABITS** Adult flies at night and is attracted to light. Larva feeds on birch and overwinters. **STATUS** Widespread and common. In Britain, more common in central and southern regions.

Blood-vein ■ *Timandra comae* Wingspan 32mm

DESCRIPTION A rather delicate moth with sharply angled fore and hindwings that are held outstretched and flat when at rest. Base colour is uniform pale buff with a distinctive dark vein-like red line that traverses the wings from the tips of the forewings, giving the moth its name. Sexes are distinguishable as male has feathered antennae. Larva is thin and brown. **SEASON** Adult flies May–Sep in 2 broods. **HABITAT** Meadows, field margins, gardens, wasteland and common land. **HABITS** Adult flies at night and is attracted to light. Larva feeds on low-growing plants such as Dock and Sorrel and overwinters. **STATUS** Widespread and locally common.

Riband Wave ■ *Idaea aversata* Wingspan 30mm

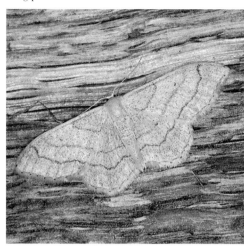

DESCRIPTION A delicate moth that occurs in 2 distinct forms. The wings in each case are yellowish-buff in colour, traversed with either a dark, broad band or 3 thin, concentric lines. At rest the wings are held flat and outstretched. Larva is slim and green to brown. **SEASON** Adult flies Jun–Oct occasionally in 2 broods. **HABITAT** Occurs in a range of habitats including grassland, woodland, heathland, hedgerows and gardens. **HABITS** Adult flies at night, is attracted to light and can be disturbed from vegetation during the day. Larva feeds on a range of low-growing plants such as Dock, Dandelion and bedstraws. **STATUS** Widespread and common.

Red Twin-spot Carpet ■ *Xanthorhoe spadicearia* Wingspan 20mm

DESCRIPTION An attractively marked moth that derives its name from a broad reddish-brown band that runs across the wings, and diagnostic dark twin spots near the tip of the forewing. Orange-buff stripes border the central band, the head and forewing shoulders being reddish-brown. Wings are held flat, forewings overlapping the hindwings. Larva is green. **SEASON** Adult flies Apr–Aug. **HABITAT** Occurs in a variety of habitats including woodland, hedgerows, downland, gardens and moorland. **HABITS** Adult flies at night and comes to light; often disturbed from vegetation during the day. Larva feeds on a variety of low-growing herbaceous plants. **STATUS** Widespread and common.

Silver Ground Carpet ■ *Xanthorhoe montanata* Wingspan 26-28mm

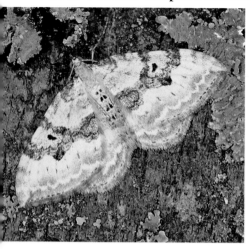

DESCRIPTION A pale moth with whitish wings and a broad, dark band that runs across the forewings, bisected centrally by the abdomen. The width of the central band is variable; the colour ranging from light to dark brown. Wings are held flat, the forewings overlapping the hindwings. Larva is green and twig-like. **SEASON** Adult flies May–Jul. **HABITAT** Occurs in a wide variety of habitats including heathland, downland, gardens, open woodland, hedgerows and scrub. **HABITS** Adult flies at night and is attracted to light; often disturbed from vegetation during the day. Larva feeds on herbaceous plants including bedstraws. **STATUS** Widespread and common.

Yellow Shell ■ *Camptogramma bilineata* Wingspan 20-24mm

DESCRIPTION This species can be variable in colour, most commonly bright yellow but ranging to dark brown. The intricate pattern of ragged, concentric brown lines and dark-bordered white lines is consistent across the range. At rest the wings are held flat and outstretched. Larva is green and twig-like. **SEASON** Adult flies Jun–Aug. **HABITAT** Occurs in a variety of habitats including woodland, hedgerows, heathland, common land and gardens. **HABITS** Adult flies from dusk onwards and comes sparingly to light; often disturbed from vegetation during the day. Larva feeds on various low-growing plants including Dock and Chickweed. **STATUS** Widespread and common.

Barred Straw ■ *Eulithis pyraliata* Wingspan 35mm

DESCRIPTION Adult has buffish-yellow wings marked with wavy dark brown lines. At rest the wings are spread wide and flat and obscure the hindwings. This diagnostic pose makes this species readily identifiable. Larva is thin and green with a dark line running along its back. **SEASON** Adult flies May–Sep. **HABITAT** Rough grassland, hedgerows, roadside verges, gardens and woodland edges where there is an abundance of the larval foodplants. **HABITS** Adult flies at night and comes to light; sometimes disturbed from vegetation by day. Larva feeds on bedstraws and Common Cleavers. **STATUS** Widespread and relatively common, scarcer in the north.

Barred Yellow ■ *Cidaria fulvata* Wingspan 20-25mm

DESCRIPTION An unmistakable moth that, as its name suggests, has yellow wings with a broad, brown central band; pale triangles adorn each wingtip. When at rest, wings are held flat and outstretched, the forewing covering the hindwing in an overall triangular shape. Larva is green and grub-like. **SEASON** Adult flies Jun–Aug. **HABITAT** Gardens, hedgerows, open woodland and common land. Directly linked to presence of the larval foodplant. **HABITS** Adult flies mainly at dusk when it is active over low vegetation; sometimes attracted to light at night. Larva feeds on wild and cultivated species of rose. **STATUS** Widespread and locally common.

Green Carpet ▪ *Colostygia pectinataria* Wingspan 22-25mm

DESCRIPTION A distinctive moth that is relatively easily identified when freshly emerged by its bright green wings. Two dark-edged, wavy white lines traverse the forewings, each terminating in a dark triangle on the leading edge and a single dark patch on the trailing edge. Colour soon fades to a lighter yellow or even whitish. Larva is mottled green. **SEASON** Adult flies May–Jul. **HABITAT** Occurs in a variety of habitats including gardens, hedgerows, heathland, downland and open woodland. **HABITS** Adult flies from dusk and is attracted to light; often disturbed from vegetation during the day. Larva feeds on bedstraws and overwinters. **STATUS** Widespread and common.

Winter Moth ▪ *Operophtera brumata* Wingspan 28-33mm

DESCRIPTION The iconic moth of the winter months. The male has rounded wings of grey-brown with dark, wavy concentric cross-lines. Wings are held flat at rest, forewings

covering the hindwings. The female is wingless and spider-like. Larva is green. **SEASON** Adult flies Nov–Feb. **HABITAT** Occurs wherever there are deciduous trees and shrubs. **HABITS** Flies at night only during the winter months and can be attracted to light. Often seen at lighted windows or in car headlights on mild nights. Mating pairs can be found on tree trunks after dark. Larva feeds on oak, birch, sallow and apple. **STATUS** Widespread and common.

Foxglove Pug ▪ *Eupithecia pulchellata* Wingspan 18-20mm

DESCRIPTION An attractive little moth and one of the easier species of pug to identify. Wings are reddish-brown and slate-grey, the 2 colours forming broad concentric bands that are edged in white. Some local colour variation and wearing can occur. Wings are held flat and outstretched at rest. Larva is green and grub-like. **SEASON** Adult flies May–Jun. **HABITAT** Occurs where larval foodplant is found including woodland, gardens, downland, moorland, clifftops and shingle beaches. **HABITS** Adult flies at night and is readily attracted to light. Larva feeds inside the flowers of Foxglove and overwinters as a pupa. **STATUS** Widespread and locally common.

Lime-speck Pug ▪ *Eupithecia centaureata* Wingspan 20-24mm

DESCRIPTION A narrow-winged moth that is predominantly white in colour, each forewing being decorated with a single dark spot on its leading edge. The markings are thought to resemble a bird dropping, affording effective camouflage. Wings are held flat and outstretched away from the body when at rest. Larva is whitish with dark markings. **SEASON** Adult flies Apr–Sep in 2 broods. **HABITAT** Wasteland, parks, gardens, hedgerows and open woodland. **HABITS** Adult flies at night and is readily attracted to light; often found at rest on flat surfaces such as walls during the day. Larva feeds on the flowers of low-growing plants. **STATUS** Widespread and fairly common.

Green Pug ■ *Pasiphila rectangulata* Wingspan 17-21mm

DESCRIPTION A beautiful pug with rounded wings that are overall green in colour with markings that comprise a pattern of concentric dark lines and bands. This species can display a wide variation in shade and can also be found in dark forms lacking any green colouration. Wings are held flat and outstretched when at rest. Larva is green with a dark dorsal line. SEASON Adult flies Jun–Jul. HABITAT Found where fruit trees occur including orchards, parks, gardens and woodland. HABITS Adult flies at night and is readily attracted to light; often found at lighted windows. Larva feeds on the flowers of fruit trees. STATUS Widespread and common.

The Magpie ■ *Abraxas grossulariata* Wingspan 38mm

DESCRIPTION The adult moth is a spectacular insect, with wings that are white overall and strikingly marked with black spots and bands of yellow; the abdomen is also boldly marked with black and yellow. The bright markings are thought to serve as a warning to birds and other predators that they are distasteful. Larva is also black, yellow and white. SEASON Adult flies Jul–Aug. HABITAT Open woodland, scrub, grassland and mature gardens. HABITS Adult flies mainly at night, sometimes on the wing during the day and can be disturbed from vegetation. Larva feeds on various shrubs. STATUS Widespread and locally common although range and numbers are in decline.

Scorched Wing ■ *Plagodis dolabraria* Wingspan 25mm

DESCRIPTION An interesting moth, its markings resemble a piece of burnt paper and give the species its name. The wings are overall orange-brown in colour, being a base of brown-buff with tightly packed, darker brown lines. The trailing edges on hind and forewings and tail of the abdomen are tinged darker or 'scorched'. Wings are held flat at rest, abdomen curved upwards. Larva is brownish and twig-like. **SEASON** Adult flies May–Jun. **HABITAT** Deciduous or mixed woodland preferring open clearings and woodland rides. **HABITS** Adult flies at night, the male can be attracted to light. Larva feeds on deciduous trees including oak, birch and sallow. **STATUS** Widespread and fairly common.

Brimstone Moth ■ *Opisthograptis luteolata* Wingspan 28-42mm

DESCRIPTION An unmistakably colourful moth with bright yellow wings, distinctively marked with a chestnut coloured patch to the tips of the forewings and a small white crescent-shaped spot. At rest the wings are held flat, forewing covering the hindwing in an overall triangular shape. Larva is green and grub-like. **SEASON** Adult flies Apr–Oct in 2 or 3 broods. **HABITAT** Hedgerows, open woodland and gardens. **HABITS** Adults fly from dusk onwards and are readily attracted to light, although they are sometimes seen during the day and mistaken for butterflies. Larva feeds on Hawthorn, Blackthorn and similar trees. **STATUS** Widespread and common.

Speckled Yellow ■ *Pseudopanthera macularia* Wingspan 26-30mm

DESCRIPTION As its name suggests, this species has bright yellow wings and a series of large dark spots or blotches. Colour can vary with some specimens being cream or almost white. Wings are held flat and outstretched at rest, the hindwings visible. Larva is green and twig-like. **SEASON** Adult flies Apr–Jul. **HABITAT** Open woodland and scrubland in lowland and upland regions. Prefers warmer climates. **HABITS** Adult flies during the day and is often mistaken for a butterfly. Larva feeds on Woodsage, White Dead-nettle and Woundwort. Overwinters as a pupa. **STATUS** Widespread and locally common. In Britain, locally common only in the south.

Purple Thorn ■ *Selenia tetralunaria* Wingspan 45-52mm

DESCRIPTION An attractive, largish moth with rather angular wings and jagged margins to the trailing edges. Underwings are purplish-brown, flushed with chestnut towards

the base and a small, pale half-moon mark close to the leading edge. Holds its wings in a diagnostic upright posture, either slightly apart at an acute angle or folded together. Larva is brown and twig-like. **SEASON** Adult flies Apr–Aug in 2 or 3 broods. **HABITAT** Deciduous woodland, gardens, heathland and parkland. **HABITS** Adult flies at night and frequently attracted to light. Larva feeds on a variety of deciduous trees. Overwinters as a pupa. **STATUS** Widespread and fairly common.

Swallow-tailed Moth ■ *Ourapteryx sambucaria* Wingspan 50-62mm

DESCRIPTION A large moth with rather angular pale yellow wings decorated with thin, brown cross lines. The hindwings are fringed brown and pointed with a short tail-streamer, recalling the Swallowtail butterfly (p.16). At rest the wings are held flat and outstretched, the hindwings visible. Larva is brown and twig-like. **SEASON** Adult flies Jun–Jul. **HABITAT** Woodland, parkland, gardens and hedgerows. **HABITS** Adult flies at night and is readily attracted to light, often found at lighted windows. Larva feeds on various shrubs including Hawthorn and ivy. Overwinters as a larva. **STATUS** Widespread and locally common. In Britain, common only in central and southern regions.

Brindled Beauty ■ *Lycia hirtaria* Wingspan 40-45mm

DESCRIPTION A well-marked species with subtly patterned wings of brown, grey and buff providing excellent camouflage when at rest. The forewings are rounded at the tip, held flat at rest and folded concealing the hindwings. Thorax and abdomen are distinctively furry. The male has feathered antennae. Larva is mottled and twig-like. **SEASON** Adult flies Mar–May. **HABITAT** Found in a wide variety of habitats where deciduous trees occur including woodland, parkland and gardens. **HABITS** Adult flies at night and is readily attracted to light. Larva feeds on various deciduous trees and shrubs. Overwinters as a pupa. **STATUS** Widespread and locally common. In Britain, common only in the south.

Peppered Moth ■ *Biston betularia* Wingspan 45-60mm

DESCRIPTION Occurs in 3 distinct forms. Most commonly seen with strikingly white wings and an exquisite mottling or 'peppering' of small dark spots and markings. Dark form has plain blackish wings; intermediate form with evenly mottled light and dark wings. At rest wings are held flat and outstretched with hindwings visible. Larva is green and twig-like. **SEASON** Adult flies May–Aug. **HABITAT** Found in a wide variety of habitats where deciduous trees occur including woodland, parkland and gardens. **HABITS** Adult flies at night and is readily attracted to light. Larva feeds on various deciduous trees and shrubs. Overwinters as a pupa. **STATUS** Widespread and common.

Light Emerald ■ *Campaea margaritata* Wingspan 40-55mm

DESCRIPTION A largish moth that when freshly emerged is a light green colour, quickly fading with age to a ghostly whitish-green. Straight, dark-edged white lines traverse the

fore and hindwings. Wings are angular and are held flat and outstretched at rest, the hindwings visible. Larva is green and twig-like. **SEASON** Adult flies Jul–Sep. **HABITAT** Woodland, gardens, parkland and hedgerows where deciduous trees and shrubs proliferate. **HABITS** Adult flies at night and is readily attracted to light; often disturbed from vegetation during the day. Larva feeds on oak, birch, beech and Hawthorn. Overwinters as a larva. **STATUS** Widespread and common.

Privet Hawk-moth ■ *Sphinx ligustri* Wingspan 100-110mm

DESCRIPTION A magnificent species and northern Europe's largest resident moth. Forewings are brown and marked with darker lines and patches; recalls tree bark. At rest, wings are held in a tent-like manner concealing the hindwings; when agitated the forewings are spread revealing pale pink-striped wings and abdomen. Larva is green with diagonal purple and white stripes and a dark 'horn' at the tail. SEASON Adult flies Jun–Jul. HABITAT Gardens, woodland rides, open countryside and commons. HABITS Adult flies at night and is attracted to light. Larva feeds on privet, lilac and ash. STATUS Widespread and locally common. In Britain, restricted to central and southern regions.

Lime Hawk-moth ■ *Mimas tiliae* Wingspan 65-80mm

DESCRIPTION A distinctively marked moth that resembles crumpled leaves. Wings are jagged at the trailing edge and typically olive-green with pinkish marbling and darker blotches, although some variation can occur. At rest wings are held flat and outstretched. Larva is green with yellow stripes and a blue 'horn' at the tail end. SEASON Adult flies May–Jun. HABITAT Deciduous woodland, parks, gardens, hedgerows and tree-lined urban avenues. HABITS Adult flies at night and is attracted to light. Larva feeds on lime, birch, elm and alder. Overwinters as a pupa. STATUS Widespread and locally common. In Britain, restricted to lowland central and southern England and Wales.

Eyed Hawk-moth ▪ *Smerinthus ocellata* Wingspan 80-90mm

DESCRIPTION A large and beautifully marked moth with striking marbled grey-brown forewings that have a slightly jagged trailing edge. The species gets its name from the

bold eye spots that adorn the hindwings and are exposed to predators as a deterrent when threatened. Larva is green with a bluish spike at the tail end. **SEASON** Adult flies May–Jul. **HABITAT** Open woodland, orchards, gardens and tree-lined waterside locations. **HABITS** Adult flies at night and can be attracted to light. Larva feeds on sallow, apple and aspen. Overwinters as a pupa. **STATUS** Widespread and locally common. In Britain, restricted to central and southern regions.

Poplar Hawk-moth ▪ *Laothoe populi* Wingspan 70-90mm

DESCRIPTION A large moth that has grey-brown forewings with a darker broad central band containing a white crescent-shaped mark. The trailing edges of fore and hindwings are jagged; overall resembles a crumpled leaf. When threatened, the hindwings project beyond the forewings exposing a reddish patch. Larva is green with pale diagonal stripes and a 'horn' at the tail end. **SEASON** Adult flies May–Aug. **HABITAT** Open woodland, waterside areas, gardens and parks. **HABITS** Adult flies at night and is readily attracted to light. Larva feeds on poplar, sallow and willow. Overwinters as a pupa. **STATUS** Widespread and fairly common.

Hummingbird Hawk-moth
■ *Macroglossum stellatarum* Wingspan 40-45mm

DESCRIPTION A medium-sized moth with a plump, furry body. Forewings are a warm grey-brown with subtle dark, wavy markings; hindwings are orange and visible when in flight. Larva is pale green with a white stripe, white spots and a bluish 'horn' at the tail end. **SEASON** Adult most common Aug–Sep but can occur outside of these months. **HABITAT** Occurs where flowers are abundant including parks, gardens, meadows and wasteland. **HABITS** A day-flying species that hovers around flowers with an audible '*hum*' and sips nectar. Larva feeds on bedstraws. **STATUS** Irregular summer migrant, numbers vary year on year. In Britain, commonest on the south coast.

Elephant Hawk-moth
■ *Deilephila elpenor* Wingspan 70mm

DESCRIPTION A large and impressive moth with rather angular and pointed forewings that are pink and olive-green in colour; hindwings are pink and black. Body is plump, furry and striped olive-green and pink. At rest wings are held flat, forewing covering the hindwing. Larva is brown or green with eyespots; resembles an elephant's trunk giving the moth its name. **SEASON** Adult flies May–Jun. **HABITAT** Woodland clearings, gardens, river valleys, meadows, commons and wasteland. **HABITS** Adult flies at night and is readily attracted to light; sometimes found feeding at flowers at dusk. Larva feeds on Willowherb and bedstraws. **STATUS** Widespread and relatively common.

Buff-tip ■ *Phalera bucephala* Wingspan 55-70mm

DESCRIPTION A distinctive and unmistakable moth that has a buff-coloured head, silver-grey wings and a large buff patch at the end of each forewing. At rest, the forewings are folded back over its body in an upright 'rolled' posture; resembles a broken twig affording excellent camouflage. Larva is boldly marked with yellow and black and covered in bristly hairs. **SEASON** Adult flies May–Jul. **HABITAT** Open woodland, hedgerows, gardens and parks. **HABITS** Adult flies at night and is attracted to light. Larva feeds on various deciduous trees including sallow, oak, birch and hazel. Overwinters as a pupa. **STATUS** Widespread and common in lowland areas.

Puss Moth

■ *Cerura vinula* Wingspan 60-70mm

DESCRIPTION A fairly large whitish moth with a furry body that is tinged with green when freshly emerged, quickly fading with age. The forewings are marked with intricate black wavy lines and buff veins. Black spots adorn the furry thorax. At rest the wings are held in a tent-like manner. Larva is green and plump with a large head and 2 antennae-like tail appendages. **SEASON** Adult flies May–Jul. **HABITAT** Open woodland, parks, gardens, hedgerows and waterside areas. **HABITS** Adult flies at night and is sometimes attracted to light. Larva feeds on poplar and sallow. Overwinters as a pupa. **STATUS** Widespread and fairly common in lowland areas.

Sallow Kitten ■ *Furcula furcula* Wingspan 35-42mm

DESCRIPTION An attractively marked moth with whitish forewings boldly marked with a broad, orange-bordered central band. The outer margins of the wings are decorated with black spots and the thorax is dark grey and orange. At rest it holds its wings in a tent-like manner. Larva is green and fat with an orange-brown 'saddle' and 2 tail appendages. SEASON Adult flies May–Aug in 2 broods. HABITAT Damp woodland, waterside areas, parks and gardens. HABITS Adult flies at night and is readily attracted to light. Larva feeds mainly on sallow. Overwinters as a pupa. STATUS Widespread and common. Restricted to lowland areas in the north.

Pale Prominent ■ *Pterostoma palpina* Wingspan 40-60mm

DESCRIPTION A largish, elongated moth with pale greyish-buff wings that are held in a tent-like manner when at rest. The protruding abdomen tip and snout, together with the jagged wing edges and overall profile, recalls a piece of broken wood affording excellent camouflage against predators. Antennae are feathered. Larva is green, plump with yellow lateral stripes. SEASON Adult flies May–Aug in 2 broods. HABITAT Damp deciduous woodland, gardens, parks and commons. HABITS Adult flies at night and is readily attracted to light. Larva feeds on poplar and sallow. Overwinters as a pupa. STATUS Widespread and common; more local in the north.

Chocolate-tip ■ *Clostera curtula* Wingspan 30-42mm

DESCRIPTION An attractive moth with grey-brown forewings marked with dark-edged white cross lines and large maroon patch to the wing-tip. At rest wings are rolled in a tight tent-like fashion. The body is furry and the head and thorax are marked with a dark brown stripe. When alarmed, exposes brown-tipped abdomen. Larva is green with a black head. **SEASON** Adult flies Apr–Sep in 2 broods. **HABITAT** Woodland, scrub, hedgerows, gardens and plantations. **HABITS** Adult flies at night and is attracted to light. Larva feeds on aspen, poplar and willow. Overwinters as a pupa. **STATUS** Widespread and common. In Britain only locally common in the south.

The Vapourer ■ *Orgyia antiqua* Wingspan 25-40mm

DESCRIPTION The male has rather plain, rounded reddish-brown forewings that are marked with a single, half-moon-shaped white spot. At rest the forewings are folded in a heart-like shape concealing the hindwings. The female has rudimentary wings and is flightless. Larva is bristly with tufts of yellow and black hair. **SEASON** Adult flies Jul–Oct in 2 broods. **HABITAT** Occurs in a wide variety of habitats including woodland, gardens, hedgerows, parks and heathland. **HABITS** Adult males can be seen flying during the day; they also fly at night and are occasionally attracted to light. Larva feeds on deciduous trees and shrubs. Overwinters as an egg. **STATUS** Widespread and common in lowland areas.

Yellow-tail ■ *Euproctis similis* Wingspan 35-45mm

DESCRIPTION The adult has white rounded wings and a rather hairy white body. The tip of the abdomen is golden yellow, giving the moth its name. The male displays one or more dark spots to the trailing edge of the forewings and has feathered antennae. Wings are held in a tent-like manner at rest. Larva is hairy with black and red markings. **SEASON** Adult flies Jul–Aug. **HABITAT** Areas of bushy vegetation including hedgerows, gardens, scrub and woodland. **HABITS** Adult flies at night and is readily attracted to light. Larva feeds on deciduous shrubs. The hairs of the adult and larva can cause skin irritation. **STATUS** Widespread and locally common.

Common Footman ■ *Eilema lurideola* Wingspan 28-30mm

DESCRIPTION A rather elongated and narrow-winged moth with lead grey forewings, distinctively marked with a yellow leading edge that narrows towards the wing-tip; hindwings are yellow. Wings are held flat at rest, one forewing tightly overlapping the other. Larva is black and spiny. **SEASON** Adult flies Jul–Aug. **HABITAT** Occurs in a variety of habitats including gardens, woodland, hedgerows, parks and coastal areas. **HABITS** Adult flies from dusk onwards and is readily attracted to light and sugar. Larva feeds on various lichens growing on trees, shrubs and rocks. Overwinters as a larva. **STATUS** Widespread and locally common.

Garden Tiger Moth ■ *Arctia caja* Wingspan 50-78mm

DESCRIPTION A large and very striking moth with an exquisite and complex pattern of brown blotches on a variable cream base colour. Hindwings are bright orange with large dark spots. Wings are held flat at rest, forewings covering the hindwings which it spreads revealing the hindwings if threatened. Larva is dark and hairy. **SEASON** Adult flies Jul–Aug. **HABITAT** Prefers open areas in a variety of habitats including gardens, woodland, hedgerows and coastal dunes. **HABITS** Adult flies at night and is readily attracted to light. Larva feeds on a variety of herbaceous plants. Overwinters as a larva. **STATUS** Widespread and common but numbers declining.

White Ermine ■ *Spilosoma lubricipeda* Wingspan 38-46mm

DESCRIPTION An unmistakable, white moth with a fluffy white head. The white forewings are decorated with a series of small black dots, variable in size and number.

Hindwings are also white with at least one central black spot. The wings are held in a tent-like manner when at rest. Larva is dark and hairy. **SEASON** Adult flies May–Jul; occasional second brood Sep–Oct. **HABITAT** Occurs in most habitats owing to the wide variety of accepted larval foodplants. **HABITS** Adult flies at night and is attracted to light. Larva feeds on a wide variety of herbaceous plants including nettles and Dock. Overwinters as a pupa. **STATUS** Widespread and common.

Ruby Tiger ■ *Phragmatobia fuliginosa* Wingspan 28-42mm

DESCRIPTION An attractive moth that has plain reddish-brown forewings marked with 2 small black dots; hindwings are red with black blotches. The head and thorax are the colour of the forewings and distinctively furry. Wings are held in a tent-like manner, forewings covering the hindwings. Larva is reddish and hairy. **SEASON** Adult flies Apr–Sep in 2 broods. **HABITAT** Waste ground, heathland, moorland, gardens and open woodland. **HABITS** Adult mainly flies at night and can be attracted to light; sometimes encountered flying on sunny days. Larva feeds on a wide variety of herbaceous plants. Overwinters as a larva. **STATUS** Widespread and common.

The Cinnabar ■ *Tyria jacobaeae* Wingspan 32-42mm

DESCRIPTION A colourful moth with sooty-black forewings marked red stripes and spots; hindwings are crimson. The wings are held flat at rest, the forewings overlapping the hindwings. Colouration acts as a warning to predators that they are distasteful. Larva is banded with yellow and black. **SEASON** Adult flies May–Jul. **HABITAT** Grassland, meadows, waste ground and commons. **HABITS** Adult flies mainly at night and can be attracted to light; sometimes seen flying in bright sunshine or when disturbed from vegetation. Larva is a gregarious feeder and large groups are often seen feeding on Common Ragwort. Overwinters as a pupa. **STATUS** Widespread and fairly common; more local in the north of its range.

Heart and Dart ■ *Agrotis exclamationis* Wingspan 30-40mm

DESCRIPTION A distinctively marked moth with brown forewings adorned with a heart-shaped blotch and tapering black line, giving the moth its name. The front of the thorax is diagnostically marked with a black band. Wings are held flat at rest with one forewing overlapping the other. Larva is brown and grey. **SEASON** Adult flies May–Jul. **HABITAT** Occurs in virtually any habitat owing to the wide variety of accepted larval foodplants. **HABITS** Adult flies at night and is readily attracted to light. Larva feeds on a wide variety of herbaceous plants. Overwinters as a larva. **STATUS** Widespread and very common; more local in the north of its range.

Large Yellow Underwing ■ *Noctua pronuba* Wingspan 45-55mm

DESCRIPTION Forewings vary from light brown with marbled dark markings to uniform dark brown; hindwings are yellow with a black submarginal band. Wings are held flat at rest, the forewings covering the hindwings. Hindwings are exposed when in flight and act as a predator deterrent. Larva is yellow to brown. **SEASON** Adult flies Jun–Sep. **HABITAT** Occurs in virtually any habitat owing to the wide variety of accepted larval foodplants. **HABITS** Adult flies at night and is readily attracted to light; commonly disturbed from vegetation by day. Larva feeds on a wide range of herbaceous plants and grasses. Overwinters as a larva. **STATUS** Widespread and extremely common.

Cabbage Moth ■ *Mamestra brassicae* Wingspan 35-50mm

DESCRIPTION A variable moth that has brown forewings with dark mottled markings, a kidney-shaped, white-bordered spot and wavy line near the trailing wing margin. Wings are held flat or in a tent-like manner when at rest. Larva is dark with a light lateral stripe. **SEASON** Adult flies mainly May–Oct in up to 3 broods but can occur in any month. **HABITAT** Occurs in most lowland habitats. **HABITS** Adult flies at night and can be attracted to light. Larva feeds on cabbages and other brassicas and often seen as a pest. Overwinters as a pupa. **STATUS** Widespread and common; more local in the north.

Bright-line Brown-eye ■ *Lacanobia oleracea* Wingspan 35-45mm

DESCRIPTION Adult has rich brown forewings with a thin, wavy submarginal white line to the outer edge; an orange kidney-shaped mark and white circular spot complete the markings that give the moth its name. Wings are held flat at rest. Larva is green and striped. **SEASON** Adult flies May–Jul; occasional second brood flies until Sep. **HABITAT** Occurs in a variety of habitats, most commonly gardens and cultivated land. **HABITS** Adult flies at night and can be attracted to light. Larva feeds on a wide variety of herbaceous and woody plants. Occasionally regarded as a pest of tomatoes. Overwinters as a pupa. **STATUS** Widespread and common.

Antler Moth ■ *Cerapteryx graminis* Wingspan 25-40mm

DESCRIPTION A moth with forewings of dull olive-brown to reddish-buff and an undulating dark submarginal band to the outer edge. A thin 'branched' white marking resembling an antler dominates the centre, but varies in intensity and sometimes almost absent. Wings are held flat at rest. Larva is green. SEASON Adult flies Jul–Sep. HABITAT Open grassland, upland moors and downland. HABITS Adult flies at night and is attracted to light; can also be seen flying by day and feeding on flowers. Larva feeds on various grasses. Overwinters as an egg. STATUS Widespread and locally common; more abundant in the north.

Pine Beauty ■ *Panolis flammea* Wingspan 30-40mm

DESCRIPTION An attractively marked moth with orange-brown forewings marbled with darker patches and 2 light spots, one round and the other kidney-shaped. Dark and lighter orange forms can occur. Head and thorax are furry. Wings are folded in a tent-like manner at rest. Larva is green with yellow stripes. SEASON Adult flies Mar–May. HABITAT Coniferous woodland and other areas where pine trees occur. HABITS Adult flies at night and is occasionally attracted to light. Larva feeds on the needles of pine trees and is regarded as a pest in managed plantations. Overwinters as a pupa. STATUS Widespread and locally common with the exception of upland areas.

Common Quaker ▪ *Orthosia cerasi* Wingspan 30-40mm

DESCRIPTION A variable moth, the round-tipped forewings ranging in colour from grey-buff to rich brown. Commonly marked with a thin, pale submarginal line to the outer edge and 2 pale rings, one round and the other kidney-shaped; markings can vary in intensity. Rests with its wings held flat, one slightly overlapping the other. Larva is green and grub-like. **SEASON** Adult flies Mar–May. **HABITAT** Deciduous woodland, gardens, hedgerows and parks. **HABITS** Adult flies at night and is readily attracted to light. Larva feeds on deciduous trees, especially oak and willow. Overwinters as a pupa. **STATUS** Widespread and common in lowland areas.

Hebrew Character ▪ *Orthosia gothica* Wingspan 30-40mm

DESCRIPTION A well-marked moth with brown forewings, marbled with purplish-grey and rich brown. A saddle-shaped, diagnostic dark marking adorns the forewing towards the leading edge. Colour variation and intensity of markings regularly occur. Wings are held flat at rest. Larva is green with a dark lateral stripe. **SEASON** Adult flies Mar–May. **HABITAT** Occurs in virtually any habitat owing to the wide variety of accepted larval foodplants and a common garden species. **HABITS** Adult flies at night and is readily attracted to light. Larva feeds on a wide variety of deciduous trees, shrubs and herbaceous plants. Overwinters as a pupa. **STATUS** Widespread and common.

The Shark ▪ *Cucullia umbratica*
Wingspan 48-59mm

DESCRIPTION A rather drab moth with grey wings that display a subtle patterning of fine lines and tiny dots, providing excellent camouflage against tree bark. The thorax has a distinctive projecting crest. The tops of the legs are furry. At rest the wings are held tightly folded around the body in a slightly upright posture. Larva is green and grub-like. **SEASON** Adult flies Jun–Jul. **HABITAT** Downland, waste ground, gardens, coastal dunes and beaches. **HABITS** Adult flies at night and can be attracted to light. Larva feeds on Sow Thistle, wild lettuce and hawkweeds. Overwinters as a pupa. **STATUS** Widespread and common; more local in the north.

Merveille du Jour ▪ *Dichonia aprilina* Wingspan 42-52mm

DESCRIPTION A stunningly beautiful and variably marked moth. In its most common form its wings are intricately decorated with a pattern of green, black and white. A series of black blotches are closely spaced forming a band that traverses the wings. Can occur in dark forms with few obvious markings. Wings are held flat at rest, one slightly overlapping the other. Larva is brightly marked and hairy. **SEASON** Adult flies Sep–Oct. **HABITAT** Mature oak woodland, gardens and parkland. **HABITS** Adult flies at night and is attracted to light. Larva feeds on oak. Overwinters as an egg. **STATUS** Widespread and fairly common; more local in the north.

The Satellite ■ *Eupsilia transversa* Wingspan 40-47mm

DESCRIPTION A winter-flying
moth with reddish-brown to
yellowish-brown wings that are
marked with a white or orange spot
around which 2 smaller 'satellite'
spots orbit, giving the moth its
name. Wings are held flat at rest,
one slightly overlapping the other.
Larva is black and grub-like.
SEASON Adult flies Sep–Apr.
HABITAT Deciduous woodland,
parks and gardens. HABITS Adult
flies at night and is readily attracted
to light. Larva is omnivorous feeding
on deciduous trees and shrubs,
more latterly on the larvae of other
species. Overwinters as an adult.
STATUS Widespread and common,
more local in the north.

Centre-barred Sallow ■ *Atethmia centrago* Wingspan 32-36mm

DESCRIPTION As its name
suggests this species has a central
broad orange-coloured band that
traverses the yellowish wings. The
tips of the forewings also display
orange margins. The underwings are
whitish, again with orange margins.
Wings are held flat at rest, one
slightly overlapping the other. Larva
is brown and grub-like. SEASON
Adult flies Aug–Sep. HABITAT
Occurs where ash grows including
woodland, hedgerows, parks, gardens
meadows and commons. HABITS
Adult flies at night and can be
attracted to light. Larva feeds on the
mature buds and unopened flowers
of ash. Overwinters as an egg.
STATUS Widespread and locally
common.

The Sycamore ▪ *Acronicta aceris* Wingspan 35-45mm

DESCRIPTION A rather understated looking moth with wings of pale to sooty-grey marked with subtle mottling and wavy dark lines. Wings are held flat or in a shallow tent-like manner at rest, one overlapping the other. In contrast, the larva is strikingly marked being covered in orange and yellow hairs and a line of black-ringed white spots along its back. **SEASON** Adult flies Jun–Aug. **HABITAT** Urban areas and parkland, also in woodland and scrub. **HABITS** Adult flies at night and is attracted to light. Larva feeds on Field Maple and Horse Chestnut. Overwinters as a pupa. **STATUS** Widespread and locally common. In Britain, common only in the southeast.

Grey Dagger

▪ *Acronicta psi* Wingspan 35-45mm

DESCRIPTION Adult has pale grey forewings that have a powdery look and are strikingly marked with dark dagger-like blotches. The wings are held in a shallow tent-like manner when at rest, wings slightly overlapping. Larva has black and orange stripes with red spots and a prominent tuft of black hairs. Can be difficult to distinguish from **Dark Dagger** *A. tridens*. **SEASON** Adult flies Jun–Aug. **HABITAT** Deciduous woodland, parks, gardens, hedgerows and other places where there are deciduous trees. **HABITS** Adult flies at night and is readily attracted to light. Larva feeds on deciduous shrubs and trees. Overwinters as a pupa. **STATUS** Widespread and common.

Old Lady ■ *Mormo maura* Wingspan 65-75mm

DESCRIPTION A large, grey-brown broad-winged moth that has a subtle and attractive wing pattern containing bands of dark brown and lilac-grey, offering superb camouflage against tree bark. The wings are slightly serrated at the tips and are held flat when at rest. Larva is greenish, plump and grub-like. **SEASON** Adult flies Jul–Sep. **HABITAT** Deciduous woodland, parks, gardens, hedgerows and other places where there are deciduous trees. **HABITS** Flies at night and is occasionally attracted to light. Habitually finds its way indoors through open windows. Larva feeds on deciduous shrubs and trees, notably Blackthorn. Overwinters as a larva. **STATUS** Widespread and locally common.

Angle Shades
■ *Phlogophora meticulosa*
Wingspan 45-55mm

DESCRIPTION A very distinctive moth that has the appearance of a crumpled leaf. Wings are dull greenish-brown and pinkish-buff and adorned with a large dark triangular mark to the leading edge. Wing-tips are acutely angled with jagged edges. Forewings are creased lengthways when at rest. Larva is green and grub-like. **SEASON** Adult flies May–Oct in at least 2 broods but can occur in any month of the year. **HABITAT** Found in a variety of habitats including woodland, gardens, commons and urban areas. **HABITS** Adult flies at night and is readily attracted to light. Larva feeds on a wide variety of herbaceous plants. Overwinters as a larva. **STATUS** Widespread and common.

Dark Arches
■ *Apamea monoglypha* Wingspan 46-54mm

DESCRIPTION A well-marked moth with mottled wings that vary in colour from medium brown to almost black. Distinctive markings include a jagged black-and-white submarginal line to the wing-tips and a subtle pale circular and kidney-shaped mark. Wing-tips are slightly serrated and have a jagged appearance. Wings are held in a shallow tent-like manner at rest. Larva is green and grub-like. **SEASON** Adult flies Jun–Oct in 2 broods. **HABITAT** Grassland, meadows, moorland, downland and commons. **HABITS** Adult flies at night and is readily attracted to light. Larva feeds on various grasses. Overwinters as a larva. **STATUS** Widespread and common to abundant.

Common Rustic
■ *Mesapamea secalis* Wingspan 28-36mm

DESCRIPTION A variable moth that ranges from light buff-brown to dark brown. Wings have a mottled appearance and a distinctive kidney-shaped mark, outlined in white. Wing-tips are slightly serrated. Wings are held flat or in a shallow tent-like manner at rest, one slightly overlapping the other. Larva is green and grub-like. **Lesser Common Rustic** M. *didyma* is almost identical and often indistinguishable in the field. **SEASON** Adult flies Jul–Aug. **HABITAT** Meadows, heathland, moorland, downland and commons. **HABITS** Adult flies at night and is readily attracted to light. Larva feeds on various grasses. Overwinters as a larva. **STATUS** Widespread and common to abundant.

Frosted Orange ▪ *Gortyna flavago* Wingspan 32-43mm

DESCRIPTION An attractively patterned moth with mottled orange and brown forewings, traversed by a diagnostic broad orange central band marked with 2 pale spots. Resembles

decaying autumn leaves. Wings are held flat or in a shallow tent-like manner at rest. Larva is greenish with black spots and grub-like. **SEASON** Adult flies Aug–Oct. **HABITAT** Open countryside, fields, commons, marshes, gardens and wasteland. **HABITS** Adult flies at night and can be attracted to light in small numbers. Larva feeds inside the stems of thistles. Overwinters as an egg. **STATUS** Widespread and locally common. In Britain, more common in central and southern regions.

Green Silver-lines ▪ *Pseudoips prasinana* Wingspan 33-40mm

DESCRIPTION A colourful moth with beautiful and delicate green wings, traversed with diagonal dark-edged silvery-white lines that form a V shape when folded together. Hindwings are a uniform pale yellow. Wings are held in a tent-like manner when at rest. Larva is green and grub-like. **SEASON** Adult flies May–Jul with an occasional second brood in autumn. **HABITAT** Occurs most frequently in deciduous woodland but also found in hedgerows, parks and gardens. **HABITS** Adult flies at night and can be attracted to light. Larva feeds on various deciduous trees. Overwinters as a pupa. **STATUS** Widespread and common; more local in the north.

Silver Y ▪ *Autographa gamma* Wingspan 35-50mm

DESCRIPTION An attractive moth with wings of mottled grey and brown adorned with a diagnostic single white marking in the shape of the letter Y. The thorax has a prominent double crest. Darker forms can occur, usually later in the season. Wings are held in a tent-like manner at rest. Larva is green and grub-like. **SEASON** Adult flies May–Oct. **HABITAT** Can be found in most habitats. **HABITS** A common migrant from southern Europe that flies both by day and night and is readily attracted to light. Larva feeds on a variety of low-growing plants. Survival over winter in northern Europe is rare. **STATUS** Widespread and often abundant.

Spectacle ▪ *Abrostola tripartita* Wingspan 33-38mm

DESCRIPTION A well-marked moth with greyish-brown forewings and a dark central band. When viewed head-on, the front of the thorax is marked with 2 adjoining circular markings resembling a pair of spectacles, giving the moth its name. Wings are held in a tent-like manner at rest revealing projecting hair tufts. Larva is green and plump. **SEASON** Adult flies May–Sep occasionally in 2 broods. **HABITAT** Gardens, disturbed ground, woodland edges, hedgerows, ditches and other places where larval foodplant occurs. **HABITS** Adult flies from dusk onwards and is readily attracted to light. Larva feeds on Common Nettle. Overwinters as a pupa. **STATUS** Widespread and locally common.

Clifden Nonpareil ■ *Catocala fraxini* Wingspan 85-110mm

DESCRIPTION A large and impressive moth that approaches Hawk-moth proportions and is an iconic prize for record collectors. The forewings are rounded and mottled grey-brown affording excellent camouflage against tree bark. The underwings are dark brown with a striking broad lilac band. Wings are held flat at rest concealing the hindwings. Larva is large and grey. **SEASON** Adult flies Sep. **HABITAT** Deciduous woodland where Aspen is present. **HABITS** An extremely rare migrant to Britain. Adult flies at night but is not strongly attracted to light; feeds on sap and rotten fruit. Larva feeds on Aspen. Overwinters as an egg. **STATUS** Widespread in Europe, rare in Britain.

Red Underwing ■ *Catocala nupta* Wingspan 70-94mm

DESCRIPTION A large moth with rounded, marbled grey and brown forewings that are a good match for tree bark. Gets its name from the striking red underwings that are marked with bold black bars; revealed as a defence mechanism against predation. Wings are folded flat at rest concealing hindwings. Larva is grey-brown. **SEASON** Adult flies Aug–Sep. **HABITAT** Damp woodland, waterside areas, gardens and parks. **HABITS** Adult flies mainly at night and is occasionally attracted to light, sometimes flies during the day. Larva feeds on poplars and willows. Overwinters as an egg. **STATUS** Widespread and common in Europe; restricted to central and southern regions in Britain.

Mother Shipton ▪ *Callistege mi* Wingspan 25-35mm

DESCRIPTION An interesting moth with patterned forewings of grey and brown, forming a shape that is said to resemble the profile of a large-nosed old hag. Its name is derived from a 15th century Yorkshire witch of the same name. Wings are held flat at rest covering the hindwings. Larva is green and grub-like. **SEASON** Adult flies May–Jun. **HABITAT** Downland, meadows, heathland, verges and woodland rides. **HABITS** Adult flies by day in bright sunshine and is often mistaken for a butterfly. Larva feeds on clover. Overwinters as a pupa. **STATUS** Locally common. In Britain more common in central and southern regions.

The Herald ▪ *Scoliopteryx libatrix* Wingspan 44-52mm

DESCRIPTION A colourful moth with wings patterned with brown and orange patches and white cross lines. The outer edges of its wing-tips are ragged giving an overall appearance of a decaying leaf. Hindwings are brown. Wings are held flat at rest and cover the hindwings.

Larva is green, plump and grub-like. **SEASON** Adult flies Jul–Nov and Mar–Jun following hibernation. **HABITAT** Damp woodland and waterside habitats where sallows occur. **HABITS** Adult flies at night and is sometimes attracted to light. Larva feeds on sallow and poplar. Overwinters as an adult and often found hibernating in outbuildings. **STATUS** Widespread and locally common, less common in the north.

The Snout ■ *Hypena proboscidalis* Wingspan 34-40mm

DESCRIPTION A distinctively shaped moth with forewings that range from buff to reddish-brown in colour and marked with dark cross lines. At rest, the wings are held flat in a broad triangular shape, its long projecting palps forming an obvious 'snout'. Larva is green and grub-like. **SEASON** Adult flies Jun–Oct in 2 broods. **HABITAT** Occurs in a wide variety of habitats where larval foodplant grows including woodland, gardens, scrub and verges. **HABITS** Adult flies from dusk onwards and is readily attracted to light; sometimes disturbed from vegetation by day. Larva feeds on Common Nettle. Overwinters as a larva. **STATUS** Widespread and common.

Silverfish ■ *Lepisma saccharina* Length 13-30mm

DESCRIPTION A distinctive, wingless insect that has a gently tapering shape from head to tail. The shape, greyish colour and metallic sheen recall a fish, giving the species its name. The tail has 3 distinctive long bristles, the head is adorned with 2 long antennae. **SEASON** Adult can be encountered at any time of the year. **HABITAT** Inhabits the moist areas of buildings such as kitchens, bathrooms and damp basements. **HABITS** A long-living insect that can reach eight years of age. Consumes starches that can be found in spilt food, paper, wallpaper and other items; often considered an unwanted guest as a result. **STATUS** Widespread and common.

Mayfly ▪ *Ephemera danica*
Length 15-25mm

DESCRIPTION A large mayfly that is often referred to by the name 'Green Drake'. The dun has dull, yellowish wings with brown veining, becoming translucent and appearing almost black at the spinner stage. The tail has 3 characteristic long filaments emanating from the tip. **SEASON** Adult flies May–Aug. **HABITAT** Chalk streams and other clean still and flowing freshwater habitats with gravel or sandy bottoms. **HABITS** Spends 1–3 years in larval form buried in a tunnel in the bottom gravel filtering organic detritus. Adults emerge in large numbers and live for a few days to mate and lay eggs. **STATUS** Common and widespread in suitable habitats.

Stonefly ▪ *Nemoura cinerea* Length 5-10mm

DESCRIPTION Adult has a flattish body and an overall brown appearance. The wings are smoky-brown and dark-veined, at rest they are folded flat and overlapping covering the

2 bristle-like tail projections. The head is adorned with 2 long antennae. Sexes are distinguishable as the male is appreciably smaller than the female. **SEASON** Adult flies May–Jul. **HABITAT** Fast-flowing streams and rivers with stony bottoms, lakes, gravel pits and ponds. **HABITS** Larva develops underwater in the riverbed and is often found clinging to submerged rocks; adult frequently found resting on bankside vegetation. **STATUS** Widespread and locally common in suitable habitats.

Field Cricket ▪ *Gryllus campestris* Length 20-26mm

DESCRIPTION A robust looking cricket with a round, broad and mainly black body marked with yellow at the base of its brownish wings; head is large and black. Sexes can be determined by the vein patterning on the wings: complex and intricate in the male, regular in female. **SEASON** Adult occurs May–Jul. **HABITAT** Dry grassland, downland and heathland. **HABITS** Adult lives in a burrow, the male often sitting at the entrance on warm evenings making a '*chirping*' song. Feeds on grasses and small insects. Overwinters as a larva. **STATUS** Widespread and locally common; less frequent in the north. In Britain, restricted to a few southern sites.

Mottled Grasshopper ▪ *Myrmeleotettix maculatus* Length 11-17mm

DESCRIPTION A small, stout grasshopper with a variable ground colour from yellowish-brown or reddish-brown to green and distinctly mottled. Sexes are distinguishable by size and antennae shape; smaller male has clubbed antennae, larger female has antennae that are swollen and relatively uniform. **SEASON** Adult occurs Jun–Oct; larva hatches in May. **HABITAT** Dry areas with sparse vegetation and free-draining soil including heathland, sand dunes and downland. **HABITS** Adult has a relatively long lifespan. Makes a characteristic rrr rrr rrr song that it delivers in 10–15 second bursts. Adult feeds on grasses. Overwinters as an egg. **STATUS** Fairly common and widespread in suitable habitats.

Common Field Grasshopper ■ *Chorthippus brunneus* Length 14-25mm

DESCRIPTION A variably coloured grasshopper that is commonly brownish though striped and mottled forms occur including buff, orange and purple shades. The body is fairly delicate and slim with sharply incurved pronotal side-keels. The underside of the body is hairy in both sexes. **SEASON** Adult occurs Jul–Oct; larva hatches in Mar–Apr. **HABITAT** Dry areas of sparse vegetation with free-draining soil including downland and coastal grassland. **HABITS** Adults are good fliers and have been known to swarm. The male makes a repetitive *sst sst* song. Feeds on grasses. Overwinters as an egg. **STATUS** Widespread and locally common in suitable habitats.

Meadow Grasshopper ■ *Chorthippus parallelus* Length 17-23mm

DESCRIPTION Green or brown in colour with a dark grey-brown stripe running along the abdomen to the eye. Colour variation can occur, some females strikingly pink to purple.

The male is much smaller than the female. Sexes can also be separated by wing length: female's wings are short and stubby, rendering them flightless; male's are longer not extending beyond abdomen. **SEASON** Adult occurs Jun–Sep; larva hatches in Apr. **HABITAT** Grassy meadows, roadside verges, marshland, heaths and moorlands. **HABITS** Adult male makes a regular *rrrr* sound when displaying. Feeds on grasses. Overwinters as an egg. **STATUS** Widespread and common.

Common Green Grasshopper ▪ *Omocestus viridulus* Length 14-24mm

DESCRIPTION A rather slim and long-winged grasshopper that is usually a vibrant green in colour but can vary from yellowish to brown. The sides of the pronotum are gently incurved. In profile, note the slightly keeled outline of the head. Wings extend beyond the abdomen in the male, shorter in the female. **SEASON** Adult occurs Jun–Oct; larva hatches in Apr. **HABITAT** Undisturbed damp grassy meadows. **HABITS** A strong flying species found in the long grass. Possesses one of the loudest songs comprising a rapid and prolonged series of *clicks* increasing in volume. Feeds on grasses. Overwinters as an egg. **STATUS** Widespread and locally common in suitable habitats.

Stripe-winged Grasshopper ▪ *Stenobothrus lineatus* Length 15-23mm

DESCRIPTION A beautifully marked grasshopper with a green body and wings. A diagnostic white comma mark decorates the forewings that are also marked with a white stripe on the anterior margin, continuing along the pronotum and around the head. The keel of the pronotum is gently incurved. **SEASON** Adult occurs Jul–Oct; larva hatches in May. **HABITAT** Dry grassland on calcareous soils, heathland and dunes. **HABITS** Males emit a characteristic song comprising a pulsating and rhythmic rasping *buzz* that increases in intensity. Feeds on grasses. Overwinters as an egg. **STATUS** Widespread and locally common. In Britain, confined to the south.

Large Marsh Grasshopper ■ *Stethophyma grossum* Length 21-36mm

DESCRIPTION A large grasshopper that is olive-green to brown in colour, beautifully marked with lime green, yellow and black. The hind legs are marked with alternate yellow and black bands. Displays a yellow stripe along the anterior margin of the forewing. The female is larger than the male. **SEASON** Adult occurs Jul–Nov; larva hatches in May.

HABITAT Wetlands including acid bogs, wet meadows, moorland and water margins. **HABITS** A strong-flying species. Makes a sharp *ticking* sound in 3 varying verses 5–10 seconds in duration. Feeds on grasses and plants. Overwinters as an egg. **STATUS** Widespread and locally common. In Britain, extremely local and restricted to the south.

Common Groundhopper ■ *Tetrix undulata* Length 8-11mm

DESCRIPTION A small grasshopper-like insect that is variable in colour but commonly marbled grey-brown to yellow-brown. Rather compact and robust looking, the head and 'shoulders' are broad with a narrowly tapering body hidden below an extended pronotum.

The hind leg joints align roughly with the extent of the pronotum, folded wings are shorter. **SEASON** Adult may be found at any time of year; larva hatches May–Jul. **HABITAT** Favours damp grassland, moorland, meadows and forest glades. **HABITS** A strong flier and swimmer. Feeds on lichens and mosses. Overwinters in adult and larval forms. **STATUS** Common and widespread in suitable habitats.

Speckled Bush-cricket ■ *Leptophyes punctatissima* Length 10-17mm

DESCRIPTION A small, bright green and rather plump bush-cricket that is speckled with black dots and marked with a brown stripe on its back. The legs are rather long and delicate-looking as are the 2 antennae. Wings are short and stubby. Female displays a large horn-shaped ovipositor making the sexes distinguishable. **SEASON** Adult occurs Jul–Oct; larva hatches in May. **HABITAT** Hedgerows, scrub, parks, gardens and woodland edges. **HABITS** Song is weak and consists of a series of *zb zb zb* sounds. Adult feeds on leaves. Overwinters as an egg deposited in the bark of trees. **STATUS** Common and widespread. In Britain, restricted to central and southern regions.

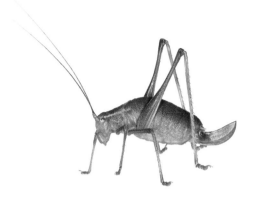

Oak Bush-cricket ■ *Meconema thalassinum* Length 12-15mm

DESCRIPTION A slender, lime-green cricket with a yellow-brown stripe running along its back. The long spindly legs and antennae give it a rather gangling appearance. Wings are of medium length. Female has a long, slightly upturned ovipositor. **SEASON** Adult occurs Jul–Nov; larva hatches in Jun. **HABITAT** Deciduous woodland and gardens with a preference for oak. **HABITS** A nocturnal species that is attracted to light. 'Song' consists of a series of drum-like beats as it strikes leaves with its hind legs. Adult feeds on small invertebrates and larvae. Overwinters as an egg deposited in the bark of trees. **STATUS** Common and widespread. In Britain, restricted to central and southern regions.

Dark Bush-cricket ■ *Pholidoptera griseoaptera* Length 15-18mm

DESCRIPTION A robust-looking cricket with a marbled dark brown body and striking yellow underside. The hind legs are rather thick-set and powerful-looking. A flightless

species, the wings are vestigial in female and no more than small flaps in male. Female has a curved, upturned ovipositor. **SEASON** Adult occurs Jun–Nov; larva hatches in April. **HABITAT** Wasteland, dry scrub, hedgerows, woodland rides and clearings. **HABITS** Adults are active by both day and night. Male makes a series of short, shrill *tsitsitsi* sounds. Adult is omnivorous and feeds on small invertebrates, larvae and vegetation. Overwinters as an egg. **STATUS** Widespread and common except in Scotland.

Grey Bush-cricket ■ *Platycleis albopunctata* Length 18-24mm

DESCRIPTION A distinctively marked cricket that is marbled grey-brown in colour with a yellow underside to the abdomen. Wings are well developed making it a strong flier. Female has a long curved ovipositor. **SEASON** Adult occurs Jun–Sep; larva hatches in May. **HABITAT** Favours dry grassland, coastal regions and areas of sparse vegetation. **HABITS** A warmth-loving species and a frequent flier on sunny days. Song consists of a series of rapid chirps *zi zi zi zib*. Adult is omnivorous and feeds on small invertebrates, larvae and vegetation. Overwinters as an egg. **STATUS** Widespread and locally common. In Britain, restricted to the south coast.

Bog Bush-cricket ■ *Metrioptera brachyptera* Length 12-18mm

DESCRIPTION A well-marked cricket with a brown body and bright green underside to the abdomen. The top of the pronotum and wings are bright green or brown, with a diagnostic light band to the hind edge. Relatively short-winged with vestigial hindwings and short forewings. Female has a long curved ovipositor. **SEASON** Adult occurs Jul–Oct; larva hatches in May. **HABITAT** Bogs, wet heathland and damp meadows. **HABITS** Adult makes a repeated and rapid series of buzzing clicks that resembles the ticking of a bicycle wheel *tsrit-tsrit-tsrit*. Feeds mainly on seeds and flowers. Overwinters as an egg. **STATUS** Widespread and locally common. Absent from Scotland except in the extreme south.

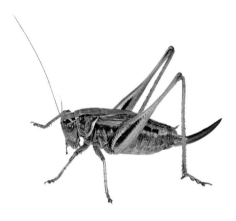

Roesel's Bush-cricket ■ *Metrioptera roeselii* Length 14-18mm

DESCRIPTION A distinctively marked cricket that is dark brown and yellow, occasionally tinted with green. The pronotum is marked all the way around the sides with a cream-coloured margin. Wings extend halfway along the abdomen. Female has curved, upturned ovipositor. **SEASON** Adult occurs Jul–Oct; larva hatches in May. **HABITAT** Occurs in a number of grassy habitats and coastal marshes. **HABITS** Mainly diurnal. Song can be heard from dusk onwards and comprises a continuous high-pitched *buzz* recalling overhead electricity lines. Feeds on grasses and small invertebrates. Overwinters as an egg. **STATUS** Common and widespread. In Britain, restricted to the south and east but distribution spreading northwards.

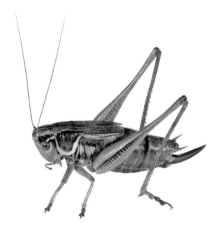

Long-winged Conehead ■ *Conocephalus discolor* Length 16-19mm

DESCRIPTION A slender, grasshopper-like cricket with a bright green body and a brown stripe along its back. The wings are brown and long, extending beyond the tip of the abdomen. The female has a long, virtually straight ovipositor. **SEASON** Adult occurs Jul–Nov; larva hatches in May. **HABITAT** Occurs in grassy habitats including damp heathland, bogs, downland, coastal reedbeds and marshes. **HABITS** Song is a series of continuous and closely spaced clicks *tsli-tsli-tsli*, inaudible to many. Omnivorous feeding on grasses, small invertebrates and larvae. Overwinters as an egg deposited in grass stems. **STATUS** Common and widespread. In Britain, restricted to the south but distribution spreading northwards.

Wartbiter ■ *Decticus verrucivorus* Wingspan 31-37mm

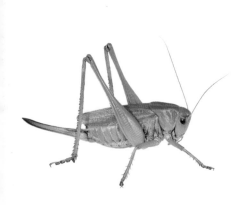

DESCRIPTION A large, robust-looking cricket whose body and wings are marbled green and brown. Wings extend beyond the abdomen, but flightless due to weight. Female has a long, slightly upturned ovipositor. **SEASON** Adult occurs Jul–Oct; larva hatches in Apr. **HABITAT** Favours bare ground and short grassland including alpine meadows, downland and heathland. **HABITS** Active by day and sings only in sunshine. Capable of delivering a painful bite. Song comprises a rapid series of clicks increasing in speed and intensity. Feeds on insects and larvae. Eggs overwinter for at least 2 seasons. **STATUS** Widespread and locally common. In Britain, now confined to a handful of sites on the south coast.

Great Green Bush-cricket ■ *Tettigonia viridissima* Length 40-46mm

DESCRIPTION The largest cricket encountered in northern Europe, the body is bright green with a brown dorsal stripe. Wings are long, green and extend beyond the tip of the abdomen. Female has a long and slightly downturned ovipositor. **SEASON** Adult occurs Jul–Oct; larva hatches in May. **HABITAT** Hedgerows, waysides and coastal scrub. **HABITS** Active by day and night and a strong flier. Song can be heard from the afternoon onwards and comprises a long and continuous series of sharp, penetrating *buzz* sounds. Omnivorous, but feeds mainly on insects and larvae. Overwinters as an egg. **STATUS** Widespread and common. In Britain, restricted to the south.

Common Hawker ■ *Aeshna juncea* Length 74mm

DESCRIPTION A large dragonfly with a dark brown body marked with paired spots to each abdominal section; blue in male and yellow to green in female. Eyes are large and blue in male, brown in female. **SEASON** Adult flies Jun–Oct. **HABITAT** Uplands, heathland, bogs, moorland, lakes, canals and ponds. **HABITS** An active and fast-flying species of upland areas that feeds on the insects it catches on the wing. Eggs are laid inside the stems of reeds and rushes. Larvae can take up to 3 years to reach adulthood. **STATUS** Common and widespread in suitable habitats. In Britain, mostly absent from the south and east.

Southern Hawker

■ *Aeshna cyanea* Length 70mm

DESCRIPTION A large dragonfly, the male has a black body with large twin green 'headlight' spots to the thorax, smaller twin spots along the abdomen, the last 2 sections being blue. Thorax sides are green and the eyes blue. The female is duller with pale green and blue markings. **SEASON** Adult flies Jun–Oct. **HABITAT** Garden ponds, lakes and canals. Shows a preference for smaller bodies of water. **HABITS** An active and fast-flying species that is often seen hunting well away from water. Feeds on insects. Female lays eggs in bankside earth and plant detritus. **STATUS** Widespread and common. In Britain, common only in central and southern regions.

Brown Hawker

■ *Aeshna grandis* Wingspan 74mm

DESCRIPTION An impressive dragonfly, easily recognizable even in flight owing to its brown body and distinctive bronze wings. The thorax is marked with 2 bold diagonal yellowish-green stripes below the wing joints. Similar markings are present on each section of the abdomen, the final 3 segments are blue in the male. **SEASON** Adult flies Jun–Oct. **HABITAT** Occurs in a wide range of habitats from small ponds and lakes to rivers and canals. **HABITS** A fast-flying species that is often seen hunting well away from water. Feeds on insects. Larvae take 2–3 years to reach maturity. **STATUS** Common and widespread. In Britain, common only in central and southern regions.

Emperor Dragonfly ■ *Anax imperator* Wingspan 78mm

DESCRIPTION A large and impressive dragonfly. The sexes are easily distinguishable: the male has a bright green thorax and sky-blue abdomen with a dark central line along its length, abdomen in the female is green to greenish-blue. **SEASON** Adult flies May–Sep. **HABITAT** Lakes, ponds and slow-flowing rivers. **HABITS** An active, wary dragonfly and a strong flier often observed hunting well away from water, feeding on insects including other dragonflies. Larval development can be rapid and ranges from 1–2 years. Eggs are laid in the stems of waterside plants. **STATUS** Common and widespread in southern regions, more local in the north. In Britain, restricted to the south.

Golden-ringed Dragonfly ■ *Cordulegaster boltonii* Wingspan 78-80mm

DESCRIPTION A large and distinctively marked dragonfly with a dark body and striking bright yellow rings to the abdomen. Eyes are bright green. Sexes can be distinguished by abdomen shape: thin and slightly club-shaped in the male, plump and straight with long ovipositor in the female. **SEASON** Adult flies May–Sep. **HABITAT** Clean, acidic fast-flowing rivers and streams. **HABITS** Adult usually only active on sunny days feeding on insects. Larval development is slow and can take 3–4 years. Larva buries itself in the bottom gravel and catches passing prey. **STATUS** Widespread and locally common in suitable habitats. In Britain, commonest in the north and west.

Northern Emerald ■ *Somatochlora arctica* Length 50mm

DESCRIPTION A relatively small dragonfly that has a green-brown metallic thorax marked with bright yellow spots. The abdomen is metallic dark green with yellow markings

on the sides of the segments. Eyes are bluish-green to yellow and the wings are suffused with yellow. Sexes distinguishable as male has calliper-like appendages at the abdomen tip. **SEASON** Adult flies Jun–Sep. **HABITAT** Moorland bogs and stagnant pools, commonly in upland areas. **HABITS** A fast-flying and agile dragonfly often seen with its abdomen slightly elevated. Eggs are laid in mud and plant detritus, the larvae taking 2–3 years to develop. **STATUS** Widespread and locally common. In Britain, restricted to northwest Scotland.

Four-spotted Chaser ■ *Libellula quadrimaculata* Length 40mm

DESCRIPTION A rather small, robust looking dragonfly, brown in colour and darkening towards the tip of the abdomen. The leading edge of each forewing is marked with 2 dark

spots, one centrally and the other towards the tip giving the species its name. **SEASON** Adult flies Apr–Sep. **HABITAT** Marshes, bogs, canals, slow-flowing rivers, shallow lakes and pools. **HABITS** A frequent and accomplished flier. The male is highly territorial and often perches around water margins fending off intruding males. Feeds on small invertebrates. Larva takes 2 years to develop. **STATUS** Widespread and locally common. In Britain, more common in central and southern regions.

Broad-bodied Chaser ■ *Libellula depressa* Length 43mm

DESCRIPTION A rather short-bodied dragonfly with a distinctive broad and flattened abdomen, the male being sky-blue with yellow spots along the edges; brown with yellow side spots in the female and immature male. Wings have dark basal markings.
SEASON Adult flies Apr–Sep.
HABITAT Small bodies of still water including ponds, canals and small lakes. **HABITS** An active flier often seen hawking for insects, can also be encountered perching on bankside vegetation. Larvae feed in the bottom mud and detritus and take 1–2 years to develop. **STATUS** Widespread and common. In Britain, common only in central and southern regions.

Black-tailed Skimmer ■ *Orthetrum cancellatum* Length 50mm

DESCRIPTION A medium-sized dragonfly with a relatively broad and flattened abdomen. Sexes are distinguishable; the male has blue eyes and a light blue abdomen with a dark tip and orange-yellow spots along the sides; the female (also immature male) is yellow-brown in colour with black lines. **SEASON** Adult flies Apr–Sep. **HABITAT** Prefers open water with bare areas of bank, most frequently large lakes, gravel pits and rivers. **HABITS** Adult feeds from perches, ambushing passing butterflies and other invertebrates. Often seen flying low to the water's surface, 'skimming'. Larval development takes up to 3 years. **STATUS** Widespread and locally common. In Britain, restricted to the south.

Keeled Skimmer ■ *Orthetrum coerulescens* Length 40mm

DESCRIPTION A smallish dragonfly that recalls the Black-tailed Skimmer (p.103), but with a more delicate, slender abdomen. The male has a brown thorax with yellowish

markings and sky-blue abdomen lacking a dark tip. The female and immature male have a yellowish to brown abdomen with a delicate dorsal line. SEASON Adult flies May–Sep. HABITAT Peat bogs, wet heathland, shallow pools and streams. HABITS Males are fiercely territorial and feed on invertebrates in dense bankside vegetation. Larvae develop in shallow water and mud, feeding on invertebrates, and take 2 years to reach maturity. STATUS Widespread and common. In Britain, locally common only in the west and southwest.

Common Darter ■ *Sympetrum striolatum* Length 36-43mm

DESCRIPTION A rather slender dragonfly, the abdomen varying in colour from deep red in mature male to orange-brown in the female and immature male. The thorax in both sexes is brown with subtle yellow markings. The adult male has a dark spot in 2 of the tail segments. SEASON Adult flies May–Nov. HABITAT Lakes, pools, garden ponds, canals and slow-flowing rivers. HABITS An active species often encountered well away from water. Uses perches such as fences and wires from which to ambush passing prey. Feeds on invertebrates. Larval development is rapid and sometimes allows for 2 broods in the same season. STATUS Widespread and common to locally abundant.

Ruddy Darter ■ *Sympetrum sanguineum* Length 35mm

DESCRIPTION A distinctive small dragonfly. The male has a bright red slender abdomen with a pronounced club-tail. The female has an ochre-yellow abdomen and thorax. Both sexes have diagnostic black legs that mark them apart from similar species. **SEASON** Adult flies Jun–Oct. **HABITAT** Prefers stagnant water such as woodland pools, weedy ponds and ditches. **HABITS** An active species that can be encountered well away from water. Perches in bushes and trees and ambushes passing invertebrates. Overwinters as an egg; larval development is rapid in the following spring. **STATUS** Common and widespread. In Britain, common only in the south but range extending northwards.

Black Darter ■ *Sympetrum danae* Length 30mm

DESCRIPTION A small dragonfly with a slender body. The male has an abdomen with a swollen club-tip that darkens with age, ultimately appearing black. The female and immature male have a yellow abdomen and a brown thorax decorated with a black triangle. Legs are black in both sexes. **SEASON** Adult flies Jul–Oct. **HABITAT** Stagnant acidic water such as peat bogs, drainage ditches and small ponds. **HABITS** A busy and skittish flier. A lover of the sun often found basking on bare ground. Feeds on invertebrates. Overwinters as an egg, larval development is rapid the following spring. **STATUS** Widespread and locally common in suitable habitats.

Downy Emerald ■ *Cordulia aenea* Length 48mm

DESCRIPTION As its name suggests, this species is green in colour, the abdomen having a distinctive metallic bronze sheen. The eyes are bright green. The thorax is hairy and the wing bases have a yellowish tinge. The male can be distinguished by the constricted abdomen that broadens towards the tail end. **SEASON** Adult flies Apr–Aug. **HABITAT** Ponds, lakes, canals, ditches and bogs. **HABITS** An active flier that is often seen flying fast and low over water. Hunts and eats invertebrates on the wing. Larval development takes 1–2 years. **STATUS** Widespread and common. In Britain, locally common with a bias towards the south.

Club-tailed Dragonfly ■ *Gomphus vulgatissimus* Length 50mm

DESCRIPTION A medium-sized dragonfly with a rather club-shaped abdomen as its name suggests. The eyes are green and distinctively widely spaced. The thorax and abdomen

in both sexes is black with bold yellow markings, turning lime green in the mature male. **SEASON** Adult flies Apr–Aug. **HABITAT** Clean, slow-flowing river systems. Sometimes occurs in ponds and lakes. **HABITS** A solitary species that is often inactive for long periods, resting on the ground or on vegetation. Feeds on invertebrates. Larval development is slow taking 3–4 years. **STATUS** Widespread and locally common. In Britain, restricted to a few river systems such as the Thames and Severn.

Banded Demoiselle

■ *Calopteryx splendens* Length 45mm

DESCRIPTION A large and distinctive species, the sexes of which are easily distinguishable. The male has a metallic blue body with smoky wings marked with a diagnostic blue 'thumbprint' mark. The female has a metallic green body with a bronze tip and greenish-brown wings. **SEASON** Adult flies May–Sep. **HABITAT** Slow-flowing lowland streams and rivers, sometimes ponds and lakes. **HABITS** Often found resting on waterside vegetation. Groups of males regularly seen flying rather lazily over water. Feeds on invertebrates. Larva lives buried in the bottom sediment and takes 2 years to develop. **STATUS** Common and widespread. In Britain, common only in England and Wales.

Beautiful Demoiselle ■ *Calopteryx virgo* Length 45mm

DESCRIPTION Similar to the Banded Demoiselle (above), the male is metallic blue but with wings that are uniformly dark and bluish-brown in colour. The female is metallic

green with a bronze tip to the abdomen and brown wings, duller than those of the Banded. **SEASON** Adult flies May–Sep. **HABITAT** Fast-flowing streams and rivers with gravel or sandy bottoms and clear, well oxygenated water. **HABITS** Adult male often found resting on a perch surveying its territory. Feeds on invertebrates it catches on the wing. Larva active at night feeding on invertebrates and takes up to 2 years to mature. **STATUS** Widespread and locally common. In Britain, common only in the south.

Azure Damselfly ■ *Coenagrion puella* Length 33mm

DESCRIPTION A rather delicate-looking damselfly, the sexes are easily distinguishable. The thorax and abdomen of the male is sky-blue in colour with prominent black markings to the thorax and black bands along the abdomen. Section 2 of the abdomen has a diagnostic U-shaped black mark. Abdomen of female is yellow-green/blue with bronze-black dorsal surface. **SEASON** Adult flies Apr–Sep. **HABITAT** Ponds, pools, lakes and marshes. **HABITS** Adults often seen mating on waterside vegetation or over pond weeds and other water plants. Larval development is swift and takes a single season, the larva overwintering. **STATUS** Widespread and common with the exception of northern Britain.

Common Blue Damselfly ■ *Enallagma cyathigerum* Length 32mm

DESCRIPTION Recalls Azure Damselfly (above) but distinguishable with care. The male's thorax has a greater concentration of blue than black and a diagnostic 'door-knob' shaped spot on segment 2. Female is typically green with a blackish dorsal surface and a spine on the underside near the abdomen tip. **SEASON** Adult flies May–Oct. **HABITAT** Vegetated still waters such as ponds, pools, lakes and canals. Occasionally flowing water. **HABITS** Often seen flying low through vegetation and open areas close to breeding sites. Mating couples form a typical 'wheel' shape during copulation. Eggs are laid within plants just below the water's surface. **STATUS** Common and widespread.

Blue-tailed Damselfly ■ *Ischnura elegans* Length 32mm

DESCRIPTION A small and distinctive damselfly. The male has a predominantly black thorax and abdomen; the penultimate abdominal section forms a broad sky-blue band giving the species its name. The female has at least 5 colour forms, typically blue, or green-brown when mature. **SEASON** Adult flies May–Sep. **HABITAT** Prefers ponds, lakes, canals, ditches and other still waters. **HABITS** A species that frequents areas close to breeding sites and can often be seen in large numbers. Adult feeds on flying insects, catching them with its legs. Larval development is rapid and 2 broods each season is common. **STATUS** Common and widespread except in northern regions.

Large Red Damselfly ■ *Pyrrhosoma nymphula* Length 35mm

DESCRIPTION Both sexes have a mainly bright red thorax and abdomen with black markings. Legs are also black. The female is generally more heavily marked than the male, appearing almost entirely black in some cases. **SEASON** Adult flies Apr–Aug. **HABITAT** Occurs in a wide variety of habitats from fast-flowing streams to lakes, ditches, marshes and even brackish water. **HABITS** One of the first species to emerge in the spring. The adult has a weak flight and can be seen in large numbers feeding on small insects in areas close to breeding grounds. Larvae are voracious feeders and mature overwinter in a single season. **STATUS** Common and widespread.

Small Red Damselfly ■ *Ceriagrion tenellum* Length 30mm

DESCRIPTION A slender and rather delicate damselfly. The adult male has an entirely red abdomen; the female is mainly bronze-black with variable degrees of red markings.

In both sexes the thorax is bronze-black on the upperside. **SEASON** Adult flies May–Sep. **HABITAT** Still acidic water including, heathland and peat bogs, forest pools, lakes and ponds. **HABITS** A weak, low-flying and inconspicuous species that is usually only observed on the wing on warm calm days close to breeding sites. Eggs are laid in waterside plants. Larval development takes 2 years. **STATUS** Widespread but only locally common in suitable habitats. In Britain, restricted to heathland bogs in the south.

White-legged Damselfly ■ *Platycnemis pennipes* Length 35mm

DESCRIPTION A slender and elegant-looking damselfly. The male is pale blue with dark markings along the centre of the abdomen and the upperside of the thorax. Legs are

extended and white with a dark central line. The female is pale green with paired black marks on each abdominal segment. **SEASON** Adult flies May–Sep. **HABITAT** Well vegetated, still and slow-flowing lowland waters such as lakes, pools, canals and rivers. **HABITS** Adults often seen sheltering in tall vegetation. The male makes a fluttering display flight to attract females. Eggs are laid into emergent waterside plants. Larval development takes a single season. **STATUS** Widespread and locally common in southern regions only.

Emerald Damselfly
▪ *Lestes sponsa* Length 38mm

DESCRIPTION A medium-sized damselfly. The male is metallic green with blue eyes, blue markings adorn the sides of the thorax and blue bands appear on the first and last 2 segments of the abdomen. The female is similar but lacks blue markings. **SEASON** Adult flies Jun–Nov. **HABITAT** Still, shallow water including ponds, lakes, moorland bogs and marshes. **HABITS** Adult holds its wings open at roughly 45 degrees when at rest. Female lays eggs in the stems of waterside plants, larvae emerge the following spring. Their short development makes them vulnerable to predation and the species does best in fish-free water. **STATUS** Widespread and locally common.

Red-eyed Damselfly ▪ *Erythromma najas* Length 35mm

DESCRIPTION A fairly robust-looking damselfly, the males having distinctive red eyes that give the species its name. Male is predominantly black with iridescent blue markings to the side of the thorax and tip of the abdomen. Females are black with yellow markings.

SEASON Adult flies May–Sep. **HABITAT** Ponds, lakes, canals and slow-flowing rivers with large expanses of floating vegetation. **HABITS** Adults frequently seen resting on floating debris and lily pads. Eggs are laid in stems of water lilies. Larvae are swift and nimble feeders and develop rapidly, emerging as adults the following spring. **STATUS** Widespread and locally common. In Britain, locally common only in central and southern England.

Hawthorn Shield Bug ▪ *Acanthosoma haemorrhoidale* Length 13-15mm

DESCRIPTION A frequently encountered and distinctive species. Overall shape is the classic 'shield' but rather slender and elongated; lateral projections of pronotum pronounced. Markings are of green and brown with a light triangle on the upper part of its back. Colouration darkens with age. **SEASON** Adult occurs all year but most commonly encountered in the warmer months. **HABITAT** Deciduous woodland, hedgerows and gardens. **HABITS** Often associated with a liking for Hawthorn berries but also feeds on the leaves of deciduous trees and shrubs. Adult overwinters, emerging in the spring to mate, the new generation being fully adult from August. **STATUS** Common and widespread; more locally common in the north.

Sloe Shield Bug ▪ *Dolycoris baccarum* Length 10-12mm

DESCRIPTION A distinctive species that is reddish-brown and greenish in colour. Banded yellowish and black abdominal segments extend beyond the outer margins of the folded wings. Antennae are banded with black and yellowish-white. Body is covered in fine hairs. **SEASON** Adult occurs all year. **HABITAT** Deciduous woodland, hedgerows and gardens. **HABITS** Feeds on the berries and leaves of deciduous shrubs and trees, with a particular liking for Blackthorn and its sloe fruits. Adult overwinters, emerging in the spring to mate, the new generation being fully adult from August. **STATUS** Common and widespread; more locally common and mainly coastal in the north.

Green Shield Bug ■ *Palomena prasina* Length 13mm

DESCRIPTION A rather oval-shaped species with subtle, rounded lateral pronotum projections. As its name suggests, it is mainly green in colour, being stippled with small black dots. Colour darkens with age and is dark bronze during hibernation. Antennae are reddish. A subtle triangle of wing membrane marks the tail end. **SEASON** Adult occurs all year. **HABITAT** Deciduous woodland, hedgerows and gardens. **HABITS** Feeds on deciduous shrubs and trees, showing a particular liking for Hazel. Adult overwinters, emerging in the spring to mate, the new generation being fully adult from September. **STATUS** Common and widespread, scarcer in the north and absent from Scotland.

Parent Bug ■ *Elasmucha grisea* Length 7-9mm

DESCRIPTION A medium-sized shieldbug that is reddish-brown in colour with an obvious black patch on the scutellum. Banded black and yellowish-white abdominal segments extend beyond the outer margins of the folded wings. **SEASON** Adult occurs all year. **HABITAT** Deciduous woodland, hedgerows, heathland and gardens where foodplant proliferates. **HABITS** Associated with birch and alder upon which the larva feeds. Adult overwinters, emerging in the spring to mate, the male dying soon after. The female survives and broods the eggs and the emerging larvae for several weeks, giving the species its name. The new generation is adult from August onwards. **STATUS** Common and widespread.

Forest Bug ▪ *Pentatoma rufipes* Length 11-14mm

DESCRIPTION A large shieldbug with a rather shiny reddish-brown body marked with a distinctive orange or cream spot at the tip of the scutellum. Banded black and orange-red

abdominal segments extend beyond the outer margins of the folded wings. Legs are orange. The lateral projections of the pronotum are pronounced and hook-shaped. **SEASON** Adult occurs Jul–Nov. **HABITAT** Deciduous woodland, orchards and gardens. **HABITS** Feeds on deciduous trees with a particular association with oak. Reaches adulthood from July and eggs are laid in August. Adult feeds on fruits, insects and caterpillars. Overwinters as a larva. **STATUS** Widespread and locally common in suitable habitats.

Capsid Bug ▪ *Campyloneura virgula* Length 4-5mm

DESCRIPTION A small but distinctively marked capsid that is mainly buff-brown in colour. Close examination reveals rather intricate markings including a bright yellow

triangular spot on the pronotum and near the tips of the forewings. Head is dark brown with long and slender antennae. **SEASON** Adult occurs Jul–Oct. **HABITAT** Deciduous woodland, hedgerows and gardens. Shows a preference for Hazel, Hawthorn and oak. **HABITS** A common predatory species that is found in a range of deciduous trees and shrubs and feeds on aphids and Red Spider Mites. Records of males are extremely rare and therefore believed to be parthenogenetic. **STATUS** Common and widespread.

Common Froghopper ■ *Philaenus spumarius* Length 6mm

DESCRIPTION A familiar, small oval-shaped froghopper notorious for its many colour forms; most commonly yellowish, brownish or black, marbled with lighter markings. Outer margin of the forewing is diagnostically convex. **SEASON** Adult occurs Jun–Sep.

HABITAT Occurs in most habitats, absent only in extreme dry or wet conditions. **HABITS** Commonly referred to as the 'Cuckoo-spit Froghopper' due to the larvae generating foam nests, commonly seen in the spring on grasses and other plants. Provides efficient protection from predators and ideal conditions for larval development. As the name suggests, this species has a striking ability to jump aiding the evasion of predators. **STATUS** Common and widespread.

Red and Black Leafhopper ■ *Cercopis vulnerata* Wingspan 40mm

DESCRIPTION A large and unmistakable froghopper with distinctive black and red markings and black head and legs. Has a slightly shiny appearance. Wing cases are well rounded at the tips and folded in a tent-like manner. **SEASON** Adult occurs Apr–Aug. **HABITAT** Occurs in a variety of wooded and open habitats such as hedgerows, meadows and woodland rides. **HABITS** Often seen resting on grass stems and other low-lying vegetation, jumping to avoid danger. Larvae are rarely observed as they feed on the roots of plants. **STATUS** Common and widespread in suitable habitats. In Britain, absent north of the Scottish Highlands.

Leafhopper ▪ *Ledra aurita* Length 13-18mm

DESCRIPTION A large and distinctive species that is greeny-grey in colour with rather bizarre looking ear-like projections on the pronotum. Head is broad and steeply angled in profile. Legs are edged with fine hairs. Wing cases are well rounded at the tips and folded in a tent-like manner. Larva is oval, pale and rather squat. **SEASON** Adult occurs May–Sep. **HABITAT** Deciduous woodland, hedgerows, gardens and parks. **HABITS** Found on lichen-covered trees, particularly oaks. This species is rarely seen owing to its excellent camouflage. Adult is capable of stridulating loudly. **STATUS** Common and widespread. In Britain, restricted to the south.

Black Bean Aphid ▪ *Aphis fabae* Length 2mm

DESCRIPTION Commonly referred to as 'Blackfly', this small aphid has a rather rounded appearance, the abdomen is short and bulbous. Dark green to black in colour, most adults are wingless and observed in large clusters. **SEASON** Adult occurs May–Sep. **HABITAT** Gardens, allotments, hedgerows and farmland. **HABITS** Most commonly seen in large numbers in the summer months on garden vegetable patches where it frequently attacks runner beans and other cultivated crops by sucking the sap and stunting growth. Overwinters as an egg laid on Spindle. Commonly eaten by ladybirds and often 'farmed' by ants which milk their honeydew. **STATUS** Widespread and seasonally abundant.

Rose Aphid ■ *Macrosiphum rosae* Length 2mm

DESCRIPTION Commonly referred to as 'Greenfly' it is a small, plump aphid that is variable in colour but usually green or pinkish. Two black horn-like projections are located near the tip of the abdomen and the legs and antennae are striped yellow and black. **SEASON** Adult occurs Jun–Oct. **HABITAT** Gardens, parks, hedgerows, open woodland and other places where the host plant occurs. **HABITS** Considered an unwelcome 'pest' by most avid gardeners, the Rose Aphid lives up to its name commonly feeding on the flowers, buds and young shoots of various roses. Found in large numbers in summer, it bears live young. **STATUS** Widespread and seasonally abundant.

Dusky Cockroach ■ *Ectobius lapponicus* Length 7-11mm

DESCRIPTION A rather flattened-looking insect with a streamlined body shape, the tough forewings covering the abdomen when folded. It is grey-brown to yellow-brown with a pale translucent margin to the pronotum and forewings. The head is dark and adorned with 2 dark coloured, long and delicate-looking antennae. The female has vestigial wings and is flightless. **SEASON** Adult occurs May–Sep. **HABITAT** Woodland edges, scrub, commons and roadside verges. **HABITS** Moves with a distinctive scuttling gait. Adults are omnivorous. Overwinters as a larva in dense grass and deep leaf litter. **STATUS** Common and widespread. In Britain, found only in the extreme south.

Earwig ▪ *Forficula auricularia* Length 13mm

DESCRIPTION A familiar insect with a rather flattened and elongated beetle-like body, chestnut-brown in colour with a slightly shiny surface. At the end of the abdomen there are 2 prominent pincer-like cerci that are curved in the male, straighter in the

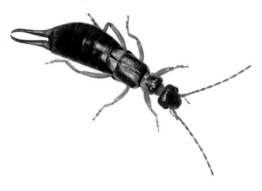

female, giving it a slightly sinister appearance. Legs and antennae are lighter yellowish-brown, the antennae clearly segmented. **SEASON** Adult occurs all year round. **HABITAT** Found in most habitats but common in woodlands, gardens, hedgerows and a familiar household guest. **HABITS** Often found under pieces of wood, large stones and leaf litter during the day. Most active at night, it feeds mainly on dead organic material. **STATUS** Common and widespread.

Alderfly ▪ *Sialis lutaria* Length 14-20mm

DESCRIPTION A stocky looking insect that has an overall black appearance. The translucent wings are heavily lined with black veins, held tightly against the body in a tent-like manner at rest. Larva is brown with a tapering abdomen fringed with gills. **SEASON** Adult flies May–Jun. **HABITAT** Banksides of well-vegetated lakes, ponds, streams and rivers with silty bottoms. **HABITS** A weak flier, adults are most commonly seen resting on the stems and leaves of bankside vegetation, living a few days only and without feeding. The larva is aquatic and lives in the bottom silt and detritus, feeding on small invertebrates. **STATUS** Widespread and common in suitable habitats.

Lacewing ■ *Chrysoperla carnea* Length 15mm

DESCRIPTION A familiar, dainty-looking insect with a pale green body, legs and antennae, and rounded transparent wings inlaid with a delicate pattern of veins. At rest, the wings are held in a tent-like manner. Antennae are long and slender. Larva is brown with 2 large pincer-like mandibles on the head. **SEASON** Adult can be found all year. **HABITAT** Occurs in most vegetated habitats. **HABITS** Adult has a weak, fluttering flight and is found amongst vegetation in the summer months. The adult hibernates and is commonly found in houses and outbuildings from autumn. Larva feeds on aphids and pupates in a cocoon constructed of its empty skins. **STATUS** Common and widespread.

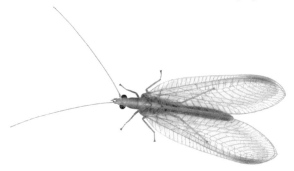

Scorpionfly ■ *Panorpa communis* Length 14mm

DESCRIPTION A distinctive and rather curious-looking insect. The body is mainly black and yellow with transparent wings that are beautifully marked with a patterning of bold black veins and blotches. The male has a scorpion-like tail to the abdomen giving the species its name. The head is downturned and with a pronounced 'beak'. Larva is caterpillar-like. **SEASON** Adult flies May–Jul. **HABITAT** Hedgerows, bramble patches, verges, woodland, nettle beds and gardens. **HABITS** Adult is active mostly at night and feeds mainly on dead insects that it steals from spider's webs, also on fruit. **STATUS** Widespread and common in the south, less frequent in the north.

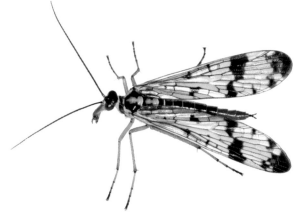

Snake Fly ▪ *Raphidia notata* Length 14mm

DESCRIPTION A rather bizarre-looking and very distinctive insect with an elongated thorax. The broad flat head projects from the base of the wings, allowing it to be raised

in a snake-like fashion giving the species its name. Wings are slender, rounded and transparent and held against the body in a tent-like manner at rest. **SEASON** Adult flies May–Jul. **HABITAT** Mature oak woodland. **HABITS** Both adult and larva can be found on the bark of mature oak trees where they feed mainly on aphids. Eggs are laid in crevices in the bark. **STATUS** Widespread and locally common, scarcer in the north.

Mottled Sedge
▪ *Glyphotaelius pellucidus*
Length 12-17mm

DESCRIPTION A moth-like species with marbled brown and white to cream wings, the pattern varying between the sexes. Wings are held against the body in a tent-like manner at rest, clearly displaying the notch at the tip of each forewing. Antennae are long and slender. **SEASON** Adult flies Apr–Oct. **HABITAT** Still waters including lakes, pools and ponds, often in woodland locations. **HABITS** Adult lays eggs attached to the underneath of waterside leaves, and the eggs fall into the water when ready to hatch. Larva constructs a case around itself from dead leaves, affording excellent camouflage from predators. **STATUS** Common and widespread in suitable habitats.

Caddisfly ■ *Phryganea grandis*
Length 18-28mm

DESCRIPTION A large species of caddisfly that is rather moth-like in appearance. Mottled reddish-brown to cream-brown in colour, the female is larger and is adorned with a dark lateral stripe allowing separation of the sexes. Wings are held in a tent-like manner at rest. **SEASON** Adult flies May–Aug. **HABITAT** Still and slow-flowing water including lakes, ponds, canals and rivers with abundant submerged vegetation. **HABITS** Adult can sometimes be attracted to light at night. Larva constructs a case around itself from plant fragments in a spiral pattern, affording excellent camouflage from predators. **STATUS** Common and widespread in suitable habitats. In Britain, mainly confined to England.

Caddisfly ■ *Limnephilus marmoratus* Length 13mm

DESCRIPTION A slender caddisfly with mottled wings of variable colour, with an attractive pattern of lighter spots and blotches. Wings are rounded at the tips and held against the body in a tent-like manner when at rest. Antennae are long and delicate. **SEASON** Adult flies May–Oct. **HABITAT** Occurs in well vegetated upland and northern lakes and tarns. **HABITS** Adult often seen resting on waterside vegetation. Larva constructs an intricate case around itself from plant fragments that provides effective camouflage. Can sometimes be seen moving around on the bottom in clear shallow water. **STATUS** Widespread but locally common in suitable habitats in upland and northern regions.

Cranefly ■ *Tipula maxima* Length 30mm

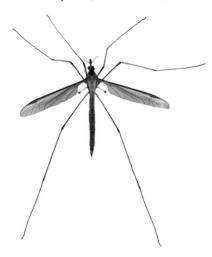

DESCRIPTION A large and beautifully marked cranefly with an impressive leg span that can reach 100mm. As with most craneflies, the abdomen is long and extended, being green-brown in colour. The wings are transparent with a brown mottling and are held outstretched at rest. **SEASON** Adult flies Apr–Aug. **HABITAT** Damp woodland and beside water in wooded areas. **HABITS** Most active at night and can be attracted to light. The life-cycle of this species revolves around water, the adults laying eggs in the damp bankside fringes, the larvae taking to the water to develop in submerged leaf litter. **STATUS** Widespread and fairly common in suitable habitats.

Leatherjacket ■ *Tipula paludosa* Length 16mm

DESCRIPTION A common species of cranefly with a classic elongated and plumpish abdomen. Colour is a uniform light to mid-brown with transparent wings and long gangly legs. Sexes are distinguishable as the wings on the female do not exceed the length of the abdomen. **SEASON** Adult flies Apr–Oct. **HABITAT** A grassland species commonly found in agricultural fields, parks, downland and gardens. **HABITS** Regularly found on garden lawns from dusk onwards and a common house guest, being readily attracted to window light. The larvae, known as 'leatherjackets', have gained a reputation with gardeners and farmers as an unwelcome pest, feeding on the roots of plants and grasses. **STATUS** Widespread and common.

Tipulid Fly ■ *Ctenophora atrata*
Length 20mm

DESCRIPTION A small wasp-like cranefly with a slender body. The smaller male varies in colour from black to yellow with spotted intermediate forms; the larger female usually has a red abdomen that tapers gently to a sharp tip. Antennae also differ; the male short and comb like, the female longer and delicately segmented. Wings are transparent and subtly veined. **SEASON** Adult flies Apr–Sep. **HABITAT** Mature deciduous woodland, orchards and other concentrations of dead and dying trees. **HABITS** Larva feeds on dead and dying wood. Species within this genus are sometimes used as a bioindicator. **STATUS** Widespread and locally common in suitable habitats.

Mosquito ■ *Culex sp* Length 8-15mm

DESCRIPTION A familiar group of small insects with rapid and buzzing flight. Difficult to see in detail without the aid of a magnifying lens; species vary but generally dark in colour with transparent wings and fly-like proportions. **SEASON** Adult flies at most times of the year, more abundant in the warmer months. **HABITAT** Can occur in any habitat, particularly abundant close to water and damp ground. **HABITS** Adult female is notorious for its habit of feeding on blood by piercing skin with its long proboscis. Male feeds on nectar. Larva develops in standing water and is small and worm-like. **STATUS** Widespread and common to locally abundant.

St Mark's Fly ■ *Bibio marci* Length 11-14mm

DESCRIPTION A slender black fly with rather large bulbous eyes and a hairy body. Wings are transparent and are folded over its abdomen when at rest, one almost completely overlapping the other. Body appears rather segmented and the legs are black. Sexes are distinguishable as female has smoky wings and smaller head and eyes. **SEASON** Adult flies Apr–Jun. **HABITAT** Occurs in grassy areas such as downland, parks, farmland and gardens. **HABITS** A rather slow and lazy flier, often low over hedgerows and grass with legs dangling below. Larvae feed on the roots of grasses and decaying vegetation. **STATUS** Widespread and locally abundant in suitable habitats.

Twin-lobed Deerfly ■ *Chrysops relictus* Wingspan 10mm

DESCRIPTION A plump and robust-looking horse fly that is beautifully patterned with bold black and yellow markings to both the abdomen and wings. The markings vary, but black lobes are always present on the second abdominal segment. The eyes are rather large and rounded and have a green iridescence. **SEASON** Adult flies May–Sep. **HABITAT** Water meadows, damp woodland and other areas of wet ground. **HABITS** Adult female can inflict a painful bite and feeds by sucking the blood of grazing animals. Male feeds on the pollen of flowers. Larvae develop in damp soil, feeding on various organic matter. **STATUS** Common and widespread in suitable habitats.

Common Horse Fly ■ *Haematopota pluvialis* Length 10mm

DESCRIPTION Sometimes referred to as the Notch-horned Cleg Fly, this species is dark grey in colour with dark grey and greenish mottling to the wings. The body is hairy and eyes beautifully mottled and iridescent. Antennae are short, black and horn-like; legs banded with dark grey and green. **SEASON** Adult flies May–Sep. **HABITAT** Moist ground such as damp woodland and pond margins. **HABITS** Adult female can inflict a painful bite and feeds by sucking the blood of grazing cattle and other mammals. Male feeds on the pollen of flowers. Eggs are laid in pond margins. Larvae develop in damp soil, feeding on small invertebrates. **STATUS** Common and widespread.

Band-eyed Brown Horse Fly ■ *Tabanus bromius* Length 14mm

DESCRIPTION This species has a grey thorax with 5 dark stripes, a hairy black abdomen with 3 rows of yellowish spots and large rounded green compound eyes. Wings are transparent with fine brown veins, the antennae short and spike-like. **SEASON** Adult flies May–Sep. **HABITAT** Meadows, downland, hill sides and other sparsely wooded spaces. **HABITS** Especially active on warm, muggy summer days. Adult female can inflict a painful bite feeding on the blood of grazing horses, cattle and other mammals including humans. Male feeds on the pollen of flowers. **STATUS** Common and widespread.

Bee Fly ■ *Bombylius major* Length 10mm

DESCRIPTION A bee mimic with a rather broad, squat and hairy body. The top of the thorax is black and shiny, the abdomen covered in thick bee-like orange-brown hair. The wings are slender and transparent with a bold black margin to the leading edge. The head is adorned with a long rigid proboscis. **SEASON** Adult flies Apr–Jun. **HABITAT** Gardens and hedgerows. **HABITS** Fond of warm sunny days where the adults can often be found using their long proboscis to feed on nectar. Adult lays its eggs in the nests of solitary bees, the emerging larvae feeding on the larvae of the hosts. **STATUS** Common and widespread.

Empid Fly ■ *Empis tessellata* Length 10mm

DESCRIPTION A rather large and slender fly with a bristly body. Drab green in colour, the wings are translucent, brown tinged and folded flat over its body at rest, one overlapping

the other. Legs are long and black, antennae short and horn-like. **SEASON** Adult flies Apr–Aug. **HABITAT** Hedgerows, woodland margins, gardens and scrub. Particularly common around Hogweed and other umbellifers. **HABITS** Adult feeds on the nectar of plants but is also an aggressive predator, feeding on other insects by piercing their bodies with its long proboscis. Males present females with a gift of a dead insect prior to mating. **STATUS** Common and widespread.

Snipe Fly ■ *Rhagio scolopaceus* Length 12mm

DESCRIPTION A slender-bodied fly with a grey-sided thorax and abdomen of alternate black and orange markings. The broad and rounded wings are transparent and shiny and are marked with an intricate patterning of dark veins and blotches. At rest they are held flat and apart, half of the abdomen visible. The dark coloured legs are long and slender. **SEASON** Adult flies May–Jul. **HABITAT** Damp vegetation mainly along woodland margins, hedgerows, moorland and meadows. **HABITS** Adult can often be found basking on damp vegetation or head-down on the trunks of trees. The carnivorous larvae can be found in leaf litter and plant detritus. **STATUS** Common and widespread.

Drone Fly ■ *Eristalis tenax* Length 12mm

DESCRIPTION A stocky hoverfly that derives its name from its resemblance to male hive bees or 'drones'. The body is rather plump and short with large eyes and slender transparent wings. The thorax and abdomen are dark with variable orange markings to the sides. **SEASON** Adult can occur at almost any time of year. **HABITAT** Hedgerows, gardens and waste ground. **HABITS** Adult visits flowers to feed on nectar and hibernates, often found in outbuildings and crevices, emerging on warm late winter days. Larva is one of the so-called 'rat-tailed maggots' inhabiting stagnant water and breathing through a long tube. **STATUS** Common and widespread.

Hoverfly ■ *Helophilus pendulus* Length 10mm

DESCRIPTION A rather slender-looking hoverfly with a black thorax marked with yellow longitudinal stripes. Each section of the abdomen is black with large bold yellow wedge-shaped markings on each side that nearly meet in the middle. Wings are slender and transparent, the legs mainly yellow and tipped black. **SEASON** Adult flies Apr–Oct. **HABITAT** Prefers areas of damp or muddy ground of woodland, field margins, hedgerows ditches and ponds. **HABITS** Adult shows a preference for warm sunny days when it is often seen basking directly in the sun. Feeds on garden flowers and wayside species such as Hogweed and Ragwort. **STATUS** Common and widespread in suitable habitats.

Hoverfly ■ *Sericomyia silentis* Length 16mm

DESCRIPTION A large robust-looking hoverfly that somewhat resembles a wasp, having a dark body and an abdomen marked with a series of 3 bold yellow bars. The eyes are large

and rounded. The wings are long, slender and transparent with a patterning of dark veins. **SEASON** Adult flies May–Nov. **HABITAT** Areas of damp, peaty soil including wet heathland, moorland and woodland. **HABITS** Adult has a rather busy flight, moving from flower to flower feeding on nectar. Larva is one of the so-called 'rat-tailed maggots' inhabiting de-oxygenated stagnant water and breathing through a long tube. **STATUS** Widespread and locally common in suitable habitats.

Hoverfly ■ *Syrphus ribesii* Length 12mm

DESCRIPTION A familiar hoverfly that has a plain olive-green to grey coloured thorax covered with short, fine orange-brown hairs. The abdomen is boldly marked with yellow and black stripes. Wings are slender and transparent with a delicate pattern of veins. **SEASON** Adult flies Apr–Oct. **HABITAT** Occurs in a wide variety of habitats including, hedgerows, gardens, woodland and waste ground. **HABITS** Males are often observed vibrating their wings in an audible hum when at rest creating a familiar backdrop of summer sound. This species can be multiple brooded during the course of a season. Larvae live on leaves feeding on aphids. **STATUS** Common and widespread.

Hoverfly ■ *Volucella bombylans* Length 14mm

DESCRIPTION A large hoverfly with big eyes that is a convincing bumblebee mimic. The thorax and abdomen are plump and furry with a variety of black and orange markings. The tip of the abdomen is either white or orange, copying the variations in bumblebee species. Wings are transparent and delicately veined and the legs are black. **SEASON** Adult flies May– Sep. **HABITAT** Commonly occurs in woodland, gardens and hedgerows. **HABITS** Adult is a regular visitor to flowers, feeding on nectar. The eggs are laid in the nests of bumblebees and wasps where the larvae scavenge on debris and occasionally the bee larvae. **STATUS** Common and widespread.

Greenbottle ■ *Lucilia caesar* Length 9mm

DESCRIPTION A rather rounded and colourful blowfly that is a bright iridescent green to bronze with large bulbous red eyes. The body is covered with stiff bristles and the wings are transparent with a pattern of subtle dark veins. One of several *Lucilia* species distinguishable only by close examination of bristle structure. **SEASON** Adult flies Apr–Oct. **HABITAT** Found in a variety of habitats including woodland margins, hedgerows and gardens. **HABITS** Adults are commonly seen on flower heads feeding on pollen and nectar or basking on vegetation. Female lays eggs on the carcasses of animals, the larvae feeding on the rotting flesh. **STATUS** Common and widespread.

Flesh Fly ■ *Sarcophaga carnaria* Length 15mm

DESCRIPTION A large fly with a coarsely bristled body, dark grey to blackish in colour with chequered markings of lighter grey on its abdomen and similarly coloured stripes to the thorax. Wings are transparent with an intricate patterning of dark veins. The large compound eyes are red and the bristly legs black. **SEASON** Adult most commonly occurs during the warmer months. **HABITAT** Can occur in almost any habitat. **HABITS** Adult seeks out carrion and carcasses on which to feed and reproduce. Produces live larvae that it deposits on decaying flesh, the larvae feeding and growing quickly before pupating. **STATUS** Common and widespread.

Tachina fera ■ *Tachina fera* Length 10mm

DESCRIPTION A large, rounded and well-marked fly. The abdomen is quite strikingly marked being mainly orange with a broad black longitudinal stripe running down the centre. The body is covered in a series of stiff bristles, the head with yellow hairs. Wings are transparent with subtle veining and a yellow flush to the base. The legs are bristly and orange, tipped with yellow at the 'feet'. **SEASON** Adult flies May–Sep. **HABITAT** Damp and well-vegetated habitats, such as damp meadows, marshes and waterside locations. **HABITS** Adult feeds on umbellifers and other waterside plants. The larvae parasitize caterpillars. **STATUS** Common and widespread in suitable habitats.

Common House-fly ■ *Musca domestica* Length 8mm

DESCRIPTION A familiar fly with a dark grey to black body, red compound eyes and orange patches on the abdomen. The body is covered in bristly hairs and the wings are transparent with a delicate vein pattern. **SEASON** Adult flies Apr–Nov. **HABITAT** The adult occurs in most habitats, particularly urban areas with concentrations of food waste. The most common of all domestic flies, it is found inside houses and other buildings where it feeds on household waste. **HABITS** Females lay large batches of eggs on decaying food upon which the hatching, maggot-like larva feeds. **STATUS** Widespread and common to abundant.

Lesser House-fly ■ *Fannia canicularis* Length 5mm

DESCRIPTION A common, slender-bodied fly that is similar to, but smaller than, Common House-fly (p.131). Thorax is brown-grey with 3 distinct black bands in males, females are more indistinctly marked. Transparent wings are veined, the fourth of which being straight and lacking the sharp bend of Common House-fly's wing pattern. **SEASON** Adult flies May–Oct. **HABITAT** Occurs in a wide variety of open habitats; most frequently encountered in buildings and around decaying corpses and dung. **HABITS** Males are often seen circling repetitively around ceilings in the centre of rooms. Larvae are maggot-like and feed on rotting flesh and excrement. **STATUS** Widespread and common.

Bluebottle ■ *Calliphora erythrocephala* Length 11mm

DESCRIPTION A familiar fly with a plump, rounded and bristly body that, as its name suggests, is a shiny dark blue in colour. The large compound eyes are reddish, the wings transparent with dark veins. **SEASON** Adult most common in the summer months. **HABITAT** Occurs in a wide variety of open habitats; most frequently encountered in buildings and around decaying corpses. **HABITS** A common household guest that makes a loud buzzing sound in flight. Attracted to meat and carrion on which it lays its eggs. The hatching larva is maggot-like and feasts on the flesh, growing and reaching pupation swiftly. **STATUS** Common and widespread.

Yellow Dung Fly ■ *Scathophaga stercoraria* Length 9mm

DESCRIPTION A brightly coloured and unmistakable fly, the male is golden-yellow and covered in fur-like hairs. The female is usually smaller and duller in colour with a green-brown tinge, lacking brightly coloured hairs on the front legs. Wings are transparent with a subtle vein pattern. **SEASON** Adult flies Mar–Oct. **HABITAT** Occurs where there are cattle such as meadows and commons. **HABITS** Large numbers of males congregate around fresh dung piles awaiting the arrival of a female. Once mated, the female lays eggs on the dung, the larvae feeding upon it and developing inside. **STATUS** Widespread and common in suitable habitats.

Picture-winged Fly ■ *Urophora cardui* Length 5mm

DESCRIPTION An attractively marked fly with a dark grey to black body, the tip of the abdomen forming a narrow funnel shape held at a slightly upturned angle. Wings are greyish-white with a bold black mark resembling a squiggly letter M along their length. **SEASON** Adult flies May–Aug. **HABITAT** Grassland and disturbed ground where the larval foodplant is found. **HABITS** The adult female lays her eggs on Creeping Thistle during the growing season. The hatched larvae burrow into the stem and form swellings called 'galls' in which the larvae develop and pupate. **STATUS** Widespread and common. In Britain common only to the south of England.

Giant Wood Wasp ■ *Urocerus gigas* Length 30mm

DESCRIPTION A large sawfly also referred to as Banded or Greater Horntail, this species is a rather formidable-looking, but harmless, wasp-like fly. The body is slender and cylindrical, marked with bands of yellow and black. The wings are transparent with a yellow tinge and patterned with dark veins. Antennae are slender and yellow. **SEASON** Adult flies May–Aug. **HABITAT** Coniferous woodland. **HABITS** The female has a long ovipositor at the tip of the abdomen that is used to penetrate decaying timber into which eggs are laid. Hatching larvae feed and develop inside the rotting wood. **STATUS** Widespread and locally common in suitable habitats.

Yellow Ophion ■ *Ophion luteus* Length 20mm

DESCRIPTION A delicate-looking insect with a slender red or orange body and dark compound eyes. The wings are transparent and patterned with subtle black veins. The legs and antennae are long and gangly-looking. The arched posture of the adult and its constantly twitching antennae gives it a slightly aggressive appearance. **SEASON** Adult flies Jul–Sep. **HABITAT** Woodland, farmland, hedgerows and gardens. **HABITS** The female lays her eggs inside the bodies of the caterpillars of noctuid moths through a hair-like ovipositor; occasionally also used in self defence as a form of 'sting' and can cause some irritation. **STATUS** Common and widespread.

Ichneumon ■ *Lissonota setosa* Length 25-30mm

DESCRIPTION A delicate-looking, slender-bodied wasp with a long abdomen that gives it an almost damselfly-like appearance. Head, thorax and abdomen are blackish in colour; the long, spindly legs are bright orange-red and wings transparent with fine dark veins and a smoky tinge. Sexes are distinguishable; the female has a long sting-like ovipositor, the male has orange-tipped, club-like antennae. **SEASON** Adult flies May–Jun. **HABITAT** Occurs wherever there is decaying wood, commonly woodland, parks, hedgerows and mature gardens. **HABITS** A species of parasitic wasp that penetrates wood with its long ovipositor laying its eggs inside the larvae of wood-boring moths and beetles. **STATUS** Common and widespread.

Ruby-tailed Wasp ■ *Chrysis ignita* Length 11mm

DESCRIPTION A small but beautifully marked insect, it is brightly coloured with a metallic green head and thorax and shiny ruby-red abdomen. Wings are narrow and transparent with a dark tinge; at rest they are folded over the abdomen, one covering the other. **SEASON** Adult flies Apr–Sep. **HABITAT** Gardens, heathland, woodland and other habitats where the host occurs. **HABITS** Female is often seen busily investigating sun-warmed surfaces such as old walls and banks, searching for the nests of the Mason Wasp. Where the adult is absent, the female parasitizes the larvae laying her eggs inside them. **STATUS** Widespread and locally common in suitable habitats.

Leaf-cutter Bee ■ *Megachile centuncularis* Length 13mm

DESCRIPTION A small bee that recalls a Honey Bee (p.139) being black with tawny-buff stripes across the abdomen and a covering of fine hairs. The underside is covered in thick orange fur. The large compound eyes are black and the antennae short and sturdy-looking. **SEASON** Adult flies Jun–Jul. **HABITAT** Found in a wide variety of habitats including woodland, grassland, heathland, farmland and gardens. **HABITS** A solitary bee, it nests in holes in dead wood, old walls and plant stems. It is characterized by the circular discs it cuts out of the leaves and uses to build cells in which their larvae live. **STATUS** Common and widespread.

Wood Ant ■ *Formica rufa* Length 10mm

DESCRIPTION A large ant, the workers of which are red and brownish-black and wingless. The abdomen is large and bulbous, the head comprises large eyes and powerful mandibles. **SEASON** Adult active Feb–Oct. **HABITAT** Coniferous and deciduous woodland and heathland. **HABITS** Forms large colonies that can number up to 500,000 individuals in impressive and conspicuous nests constructed from dead leaves, twigs and pine needles. Adults feed on aphid honeydew and insects and spray formic acid from their abdomens as a form of defence. Winged males are sometimes seen on warm days searching for a mate. **STATUS** Widespread and locally common. In Britain common only in England and Wales.

Black Garden Ant ▪ *Lasius niger* Length 3mm

DESCRIPTION Our most familiar and common ant, the majority and most frequently encountered form comprises the uniformly dark brown and wingless workers. **SEASON** Adult active Feb–Oct. **HABITAT** Can occur in a number of habitats and commonly found in gardens where it nests in dry ground, under paving slabs and bricks; a frequent house guest. **HABITS** Forms large colonies that typically number 5–7,000 individuals. Adult regularly seen 'milking' aphids for honeydew, also feeds on other insects. During hot summer days swarms of winged ants appear, comprising males and immature queens that mate on the wing, the queens dispersing to form new colonies. **STATUS** Common and widespread.

Red Ant ▪ *Myrmica rubra* Length 4mm

DESCRIPTION A common ant similar in size and appearance to the Black Garden Ant, but distinctively reddish-yellow in colour. **SEASON** Adult active throughout the year, most frequently encountered Feb–Oct. **HABITAT** Can occur in a number of habitats and commonly found in gardens and meadows nesting in dry ground and under rocks and decaying wood. **HABITS** An aggressive ant that is capable of inflicting a painful sting. Forms colonies that number thousands of individuals. During hot summer days, swarms of winged ants appear, comprising males and immature queens that mate on the wing, the queens dispersing to form new colonies. **STATUS** Common and widespread.

Hornet ■ *Vespa crabro* Length 25-35mm

DESCRIPTION A large and impressive insect with a plump and boldly marked abdomen of alternating yellow and brown-black stripes. The head is yellow and black with dark compound eyes, the thorax brown. Wings are transparent with a brown smoky tinge. **SEASON** Adult most frequently encountered Jun–Sep. **HABITAT** Woodland, gardens and parks. **HABITS** New nests are constructed every spring from wood pulp, initiated by the queen after her emergence from hibernation. A colony grows steadily reaching a peak in late summer. Feeds on other insects and can deliver a powerful sting. **STATUS** Common and widespread. In Britain, common only in the south but range extending northwards.

German Wasp ■ *Vespula germanica* Length 13-18mm

DESCRIPTION Typical wasp with a large abdomen boldly marked with yellow and black. Similar to the Common Wasp (p.139) but distinguishable by studying the head which has 3 small dark spots located between the eyes. **SEASON** Adult flies Apr–Oct. **HABITAT** Occurs in most habitats. **HABITS** Establishes large colonies in nests made from wood pulp that are generally located within, or close to the ground. Nests are started by a single queen emerging from hibernation and can grow over the course of a season to number 3,000 individuals. Adults feed on nectar and fruit and collect insects and caterpillars to feed their larvae. **STATUS** Common and widespread.

Common Wasp ▪ *Vespula vulgaris* Length 12-17mm

DESCRIPTION Typical wasp with a large plump abdomen boldly marked with yellow and black. Distinguishable from the similar German Wasp (p.138) by studying the head that

has a bold black anchor-shaped mark located between the eyes. **SEASON** Adult flies Apr–Oct. **HABITAT** Occurs in most habitats. **HABITS** Establishes large colonies in nests made from wood pulp, commonly below ground but also favours wall cavities and lofts. Nests are started by a single queen emerging from hibernation and can grow over the course of a season to number 5,000 individuals. Adults feed on nectar and fruit and feed their larvae with insects and caterpillars. **STATUS** Common and widespread.

Honey Bee ▪ *Apis mellifera* Length 12mm

DESCRIPTION A familiar and iconic insect, the abdomen classically marked with alternating black or brown and yellow stripes. Wings are transparent. **SEASON** Adult

all year but most commonly encountered May–Sep. **HABITAT** Commonly kept in hives and farmed for their honey. Seeks woodland areas in the wild and nests in tree holes. **HABITS** Live in large colonies that typically number 20,000 workers around a single queen. Best known for their construction of honeycomb cells in which their honey is stored and larva raised. The classic pollinator, the workers are often seen on flowers collecting nectar. **STATUS** Common and widespread but numbers have declined over recent years.

Red-tailed Bumblebee ■ *Bombus lapidarius* Length 11-22mm

DESCRIPTION One of the most common and easily recognizable species of bumblebee with its predominantly black furry body and bright orange tail. Wings are modest in size and transparent. Workers range from 11–16mm in size, the queen is much larger at 20–22mm and the male is 14–16mm. **SEASON** Adult flies May–Sep. **HABITAT** Hedgerows, heathland, field margins and a common garden species. **HABITS** Forms colonies that number between 100 and 200 individuals, nesting underground typically at the base of walls. Queens are often encountered in houses searching for nesting sites in the spring. Flies with an audible *buzzing* sound; a frequent visitor to flowers and an important pollinator. **STATUS** Common and widespread.

Common Carder Bee ■ *Bombus pascuorum* Length 13mm

DESCRIPTION Typical plump hairy body. The thorax is entirely ginger in colour, the abdomen being black and variably ginger, red or greyish. The male can be distinguished

by the longer antennae. **SEASON** Adult most commonly encountered May–Nov. **HABITAT** Grassland, heathland, woodland margins and gardens. **HABITS** Forms colonies that number up to 200 individuals, nesting in holes and cavities such as old birds nests and rodent burrows. The hibernating queen emerges in spring seeking a suitable nest site, laying eggs that hatch as workers. Derives its name from its habit of combing material (carding) to cover the larval cells. **STATUS** Common and widespread.

White-tailed Bumblebee ▪ *Bombus lucorum* Length 11-20mm

DESCRIPTION A classic bumblebee with a stout and hairy body, the thorax and abdomen being coloured black and yellow. Derives its name from the white patch of hair at the end of the abdomen. **SEASON** Adult flies Mar–Oct. **HABITAT** Grassland, heathland, woodland margins, farmland and gardens. **HABITS** Forms colonies; the hibernating queen emerges in the spring seeking suitable nest sites in cavities such as old birds nests and rodent holes. Lays eggs that hatch as workers, males appearing later to mate with prospective queens which then hibernate; the males, old queen and remaining colony dying. **STATUS** Common and widespread although numbers thought to be in decline.

Green Tiger Beetle ▪ *Cicindela campestris* Length 14mm

DESCRIPTION A large beetle with an iridescent green coloured body distinctively marked with yellowish-cream coloured spots. The blackish eyes are well separated and are situated on either side of the head; the legs are brown and covered in fine white hairs. **SEASON** Adult occurs Apr–Oct. **HABITAT** Areas of dry, bare ground with sparse vegetation including sandy heathland, hillsides and coastal areas. **HABITS** A ground-dwelling beetle, the adult often encountered in warm sunny spots. Both adult and larvae feed on other insects, the larvae digging burrows from which they ambush their prey. Adult will fly if disturbed. **STATUS** Widespread and locally common in suitable habitats.

Ground Beetle ■ *Pterostichus madidus* Length 14mm

DESCRIPTION A large shiny and wingless black beetle with black eyes and legs; elytra are fused together and finely grooved. Distinguishable from similar species by the rounded pronotum. A second, red-legged form also exists (var. *concinnus*). **SEASON** Adult occurs May–Nov. **HABITAT** Found in a wide variety of habitats including woodland, grassland, farmland and gardens. **HABITS** A common species of ground beetle that is frequently encountered under stones, logs, loose tree bark and grassy tussocks. Adults are mainly predatory feeding on other invertebrates, notably caterpillars and slugs but also on fruit. Some adults overwinter to breed again the following year. **STATUS** Common and widespread.

Violet Ground Beetle ■ *Carabus violaceus* Length 20-30mm

DESCRIPTION A large and unmistakable beetle, being shiny and black in colour, the edges of the smooth oval elytra and thorax being edged in violet or indigo, giving the

species its name. The legs and antennae are long and black. **SEASON** Adults can be found all year but most commonly seen in the warmer months. **HABITAT** Woodland, hedgerows, parks and gardens. **HABITS** A ground-dwelling beetle, it can be found hiding under stones and logs during the daytime, becoming active at night. A carnivorous species, it feeds on slugs, snails, worms and insects. Adult hibernates over the winter months. **STATUS** Common and widespread.

Sexton Beetle ■ *Nicrophorus vespilloides* Length 16mm

DESCRIPTION A distinctively marked beetle with a shiny black body and 4 vivid red, splash-like blotches on the elytra. Wing cases are squarish and shorter than the abdomen. Antennae and legs are black. **SEASON** Adult most active May–Oct. **HABITAT** Open woodland. **HABITS** A burying beetle that has a reproductive cycle that revolves around carrion. Adults seek out dead mammals and birds on the ground and excavate the soil beneath the carcass, covering the body with the resulting spoil. This is performed by both sexes. Eggs are laid in the burial chamber, the larvae consuming the corpse once hatched. **STATUS** Common and widespread in suitable habitats.

Devil's Coach-horse ■ *Staphylinus olens* Length 24mm

DESCRIPTION An easily recognizable species with a long slender body, matt black in colour. The head is rather ant-like and has a powerful pair of pincer-like jaws. Elytra are short leaving the abdomen exposed. **SEASON** Adult occurs Apr–Oct. **HABITAT** Woodland, hedgerows and gardens. **HABITS** A 'rove' beetle that can be found hiding under stones and logs during the day. Emerges at night to feed on invertebrates using its pincer-like jaws to crush its prey. If threatened, it takes on an aggressive posture, raising its abdomen like a scorpion and opening its jaws. Can emit a foul smell from glands in the abdomen. **STATUS** Common and widespread.

Lesser Stag Beetle ◾ *Dorcus parallelipipedus* Length 28mm

DESCRIPTION A large and impressive beetle that recalls its larger cousin the Stag Beetle (below), but more modest in size with smaller jaws. Body is broad and robust-looking and dull black in colour. The male has larger jaws than the female and lobed antennae. **SEASON** Adult occurs May–Sep. **HABITAT** Woodland, hedgerows, parks and wooded gardens **HABITS** Adults are active fliers, using this as a method of dispersal and sometimes come to light at night. Strongly associated with decaying timber, particularly ash, apple and beech, the larvae living within and feeding upon it. **STATUS** Common and widespread in suitable habitats. In Britain, common only to southern and central regions.

Stag Beetle ◾ *Lucanus cervus* Length 40mm

DESCRIPTION A large and arguably the region's most impressive beetle. The wide head and thorax are black in colour, the oval-shaped elytra reddish-brown. The male has

distinctive large reddish-brown antler-like jaws; the female is antler-less. Legs and antennae are black. **SEASON** Adult occurs May–Jul. **HABITAT** Woodland, hedgerows, and gardens. **HABITS** Can sometimes be found in numbers on warm and muggy early summer evenings, the males seeking females with which to mate. Rival males will fight each other for mating rights, clamping their large jaws around each other in an often impressive display. Larvae live and feed within buried decaying wood. **STATUS** Widespread and locally common. In Britain, restricted to the south.

Rhinocerous Beetle ■ *Sinodendron cylindricum* Length 15mm

DESCRIPTION A small but impressive and robust-looking beetle, being rather cylindrical in shape and shiny black in colour, elytra and thorax pitted. Its name is derived from the male's horn-like head projection. **SEASON** Adult occurs May–Sep. **HABITAT** Mature beech and oak woodland. **HABITS** Usually only encountered in small numbers, the adult is sometimes to be found under rotting wood. Feeds on tree sap. The female excavates a series of burrows in decaying wood within which she lays her eggs. The male has been observed protecting the burrow entrance during excavation. The hatching larvae feed and develop inside the decaying wood. **STATUS** Widespread and locally common in suitable habitats.

Rose Chafer ■ *Cetonia aurata* Length 17mm

DESCRIPTION A beautifully marked insect with a rounded, somewhat squat body. The elytra are strikingly iridescent green to bronze in colour with variable white markings that recall small cracks. The scutellum is notably triangular. **SEASON** Adult occurs Apr–Sep. **HABITAT** Woodland edges, hedgerows, meadows, parks and gardens. **HABITS** A fast flyer, the adult feeds on pollen, nectar and flowers and is particularly associated with roses, frequently seen as a garden pest as a result. Larvae live and feed on decaying wood and in compost piles where they take 2 years to reach pupation. **STATUS** Widespread and locally common but numbers thought to be in decline.

Cockchafer ■ *Melolontha melolontha* Length 35mm

DESCRIPTION A large, familiar, oval-shaped beetle with reddish-brown, ribbed and downy elytra and a pointed tip to the abdomen. Head and pronotum are blackish, antennae fan-like. **SEASON** Adult occurs May–Jun. **HABITAT** Hedgerows, open woodland, farmland and gardens. **HABITS** Sometimes referred to as 'May Bugs' due to the month in which they appear. Adults often seen swarming around the tops of trees and bushes feeding on leaves, particularly oak. Attracted to light at night and often seen at lighted windows. Larvae live and develop in the soil feeding on roots and take at least 2 years to reach pupation. **STATUS** Common and widespread except in the extreme north.

Soldier Beetle ■ *Cantharis rustica* Length 13mm

DESCRIPTION A long, slender beetle with shiny black elytra that are slightly ridged and covered in fine hairs. The pronotum is red and distinctively marked with a central dark spot. The legs are predominantly red, the antennae long and segmented. **SEASON** Adult occurs May–Jul. **HABITAT** Occurs in a wide variety of habitats including hedgerows, verges, meadows and open woodland. **HABITS** Adults are often observed on warm days on the flowers of umbellifers, thistles and other plants where they ambush other small invertebrates. Larvae are ground-dwelling and also feed on invertebrates. **STATUS** Common and widespread. In Britain, common only in England and Wales.

Glow-worm ■ *Lampyris noctiluca* Length 14mm

DESCRIPTION The male is winged with slender elongated brown elytra, clearer pronotum and central brown spot. The larviform female is wingless and larger at up to 25mm. **SEASON** Adult occurs May–Aug. **HABITAT** Grassland, meadows and open woodland especially on calcareous soils. **HABITS** Active in low-growing vegetation at night, hiding under stones and logs during daylight hours. The wingless females attract males by grasping on to the top of grass stems from dusk and emitting a greenish-yellow light from the rear segments of their abdomens. Larvae feed on invertebrates such as slugs and take 2–3 years to reach pupation. **STATUS** Widespread and locally common except for the far north.

Click Beetle ■ *Athous haemorrhoidalis* Length 14mm

DESCRIPTION A commonly encountered slender beetle. Elytra are visibly hairy and reddish-brown in colour; head and thorax darker to black. Antennae are long and straight. **SEASON** Adult occurs May–Aug. **HABITAT** Woodland, scrub, and hedgerows; associated with deciduous shrubs, most noteably Hazel. **HABITS** Derives its name from the audible click it makes if laid on its back: it flexes its body in a rapid snap and bounces into the air. This is thought to be a defence mechanism against predation. Larvae feed on plant roots; they are occasionally discovered by gardeners when digging amongst the roots of shrubs such as Hazel. **STATUS** Widespread and common in lowland areas.

Cardinal Beetle ■ *Pyrochroa serraticornis* Length 14mm

DESCRIPTION A flat-bodied beetle that derives its name from its bright scarlet colouration, consistent in this species for the head, thorax and abdomen. The bright colour protects the beetle from predation, acting as a warning to predators that they may be toxic. The legs and long, slender antennae are black. **SEASON** Adult occurs May–Jul. **HABITAT** Woodland margins, hedgerows and gardens. **HABITS** Most active on warm sunny days; often observed on flowers and leaves hunting for insects on which it feeds. Lives in rotting wood and under the peeling bark of deciduous trees. **STATUS** Widespread and locally common. Absent from Scotland.

Bloody-nosed Beetle ■ *Timarcha tenebricosa* Length 20mm

DESCRIPTION A plump, rounded beetle that is black in colour with a subtle blue metallic sheen. Elytra are smooth and retain a central line suggesting they separate but are in fact fused. Legs are robust and black, antennae thick, well segmented and black. **SEASON** Adult occurs Apr–Jun. **HABITAT** Grassland, woodland margins and heathland. **HABITS** A flightless beetle with a rather ponderous movement, often seen plodding along paths and through short grass. So named as it produces a drop of red blood-like fluid from its mouth when threatened. Larvae are blue-black in colour and feed on bedstraws; overwinter as pupae. **STATUS** Widespread and locally common.

7 Spot Ladybird ■ *Coccinella 7-punctata* Length 6mm

DESCRIPTION The classic ladybird species; small and rounded with bright reddish-orange elytra marked with seven bold black spots. The head and thorax are black with white blotches. **SEASON** Adult can be found all year; active Mar–Oct. **HABITAT** Occurs in a wide variety of habitats including woodland, hedgerows, heathland, farmland and gardens. **HABITS** A familiar ladybird that turns up wherever there are aphids to feed on, making them a welcome sight for gardeners. A migratory insect, large numbers from the continent boost British populations each spring. Often hibernates in large clusters in cavities such as hollow plant stems. **STATUS** Widespread and often abundant.

Harlequin Ladybird ■ *Harmonia axyridis* Length 9mm

DESCRIPTION A small, rounded beetle commonly deep red in colour, but can occur in a variety of colour forms from yellow to black, giving the species its name. Elytra are marked with a variable number of black spots. **SEASON** Adult active Apr–Sep. **HABITAT** Occurs in a wide variety of habitats including woodland, farmland and gardens. **HABITS** Introduced to Europe from Asia as an agricultural pest control. Releases a foul-smelling fluid if threatened. Feeds on aphids but also predates other ladybird species. Adults hibernate in large numbers in crevices and sometimes emerge on sunny winter days. **STATUS** Common and widespread in the south, range extending northwards.

Wasp Beetle ■ *Clytus arietis* Length 16mm

DESCRIPTION A slender, longhorn beetle with an elongated abdomen. As its name suggests, it is strikingly marked with yellow and black markings resembling a wasp in appearance. Legs are long, slender and reddish-brown in colour, the antennae brown and black-tipped. **SEASON** Adult occurs May–Jul. **HABITAT** Hedgerows, woodland margins and gardens. **HABITS** Adult also recalls a wasp in behaviour, often seen flying busily on warm sunny days visiting flowers and feeding on pollen. Emits a wasp-like *buzz* when threatened, but lacks the sting in the tail. Breeds in the decaying wood of deciduous trees and frequently observed resting on leaves in low vegetation. **STATUS** Common and widespread.

Longhorn Beetle ■ *Strangalia maculata* Length 16mm

DESCRIPTION Slender, rather streamlined looking beetle with long delicate antennae. Elytra are yellow to copper in colour and marked with variable black blotches and bands. Legs are similarly coloured with a mixture of yellow and black. **SEASON** Adult occurs Jun–Aug. **HABITAT** Deciduous woodland, hedgerows, heathland and gardens. **HABITS** Busy fliers, adults are particularly active on warm sunny days where they can often be seen feeding on the heads of flowers, typically umbellifers such as Hogweed. Its wasp-like body colouration is thought to deter predators. Larvae feed on a wide variety of deciduous trees and can take 2–3 years to reach maturity. **STATUS** Common and widespread.

Longhorn Beetle ▪ *Rhagium mordax* Length 21mm

DESCRIPTION A rather robust-looking longhorn beetle with relatively short and sturdy antennae and black eyes. The head, thorax and elytra are downy and coloured a uniform dull sandy-yellow with brown and black mottling and an eye-like spot to each wing-case. Legs are sandy-yellow. **SEASON** Adult occurs May–Jul. **HABITAT** Oak woodland, hedgerows and mature rural gardens. **HABITS** A rather slow-moving beetle that readily takes flight if alarmed. Adults are often encountered on the flower heads of umbellifers and hawthorn, where they feed on pollen and nectar. Larvae feed on the rotting wood of deciduous trees. **STATUS** Widespread and locally common in suitable habitats.

Dor Beetle ▪ *Geotrupes stercorarius* Length 15-20mm

DESCRIPTION Has a classic oval beetle shape with a domed body; shiny black in colour with a blue sheen. Elytra are smooth and subtly grooved. The black legs are short and thick with heavily spiked margins. Antennae are stubby and black. **SEASON** Adult occurs May–Jul. **HABITAT** Grazed grassland, commons and open woodland. **HABITS** Adults are strong fliers and capable of travelling large distances. Feeds on the dung of grazing animals. Digs burrows in the soil beneath cowpats with several chambers where the female lays her eggs, storing a piece of dung within each on which the emerging larvae feed. **STATUS** Widespread and locally common in suitable habitats.

Tortoise Beetle ▪ *Cassida viridis* Length 7mm

DESCRIPTION A small oval-shaped beetle with a flattened body, the sections seemingly without definitive separation. Leaf green in colour, the thorax and elytra have subtly pitted surfaces; underside is black. The legs and delicate antennae are brown. **SEASON** Adult occurs Apr–Oct. **HABITAT** Low vegetation in meadows and hedgerows. **HABITS** Adults can usually be found clamped on to the surface of leaves making themselves difficult for predators to prise away; colouration also provides effective camouflage. Larvae are spiny and impale cast skins and excrement on their spines as protection against predators and parasites. **STATUS** Widespread and fairly common in suitable habitats.

Poplar Leaf Beetle ▪ *Chrysomela populi* Length 10mm

DESCRIPTION A rounded, broad-shouldered beetle that bears a striking resemblance to a ladybird. Elytra are a shiny bright red colour having a smooth, subtly pitted surface; the head is black. Legs and antennae are rather short and black in colour. **SEASON** Adult occurs Apr–Aug. **HABITAT** Associated with poplars and willows and found where these occur including waterside locations and damp woodland. **HABITS** Adult and larvae feed on the leaves of poplars and willows. Adult emits a strong-smelling fluid when threatened. There are 2–3 generations per year, the last generation overwintering hidden in the leaf litter. **STATUS** Widespread and common in suitable habitats.

Mint Leaf Beetle ■ *Chrysolina menthastri* Length 9mm

DESCRIPTION A typical oval-shaped beetle with a domed body. Head, thorax and elytra are a bright, metallic, bronze-green colour with delicately pitted surfaces. Legs are of similar colouration; antennae are short, segmented and thread-like. **SEASON** Adult occurs May–Aug. **HABITAT** Damp meadows, waterside locations, hedgerows and gardens. **HABITS** As its name suggests, this distinctive species is associated with various species of mint, sometimes hemp-nettles, being the food plant for both the adult and larva. Found where the food plant occurs in boggy, waterside habitats as a result. **STATUS** Widespread and locally common in suitable habitats; less common in the north.

Green Dock Beetle ■ *Gastrophysa viridula* Length 4-7mm

DESCRIPTION A small rounded beetle with a green metallic body and an iridescence that can vary with the light. The stout legs are similarly coloured; antennae are serrated and of medium length. **SEASON** Adult occurs May–Aug. **HABITAT** Verges, field margins, water meadows and damp grassland where Dock is present. **HABITS** Strongly associated with Dock, the adult and larval food plant, and often to be found munching on its leaves. During the mating season the female's abdomen becomes enlarged and bulbous. Eggs are laid on the underside of Dock leaves. There are 2–4 broods per year, the final brood hibernating as adults. **STATUS** Widespread and common in suitable habitats.

Lily Beetle ■ *Lilioceris lilii* Length 8mm

DESCRIPTION Adult has oval-shaped elytra and slim thorax, both shiny bright red in colour recalling the Cardinal Beetle (p.148). The head, antennae, legs and underside are black. **SEASON** Adult occurs May–Aug. **HABITAT** Associated with lilies and a common garden species. **HABITS** Notorious garden beetle and unpopular with many gardeners owing to their liking for lily buds and leaves on which the adults and larvae feed. The female lays eggs on the underside of leaves, the hatching larvae feasting upon the plant before pupating in the soil. Overwinters as an adult, emerging in the spring to breed. **STATUS** Common and widespread. In Britain, common only in England and Wales.

Weevil ■ *Phyllobius pomaceus* Length 9mm

DESCRIPTION A lozenge-shaped beetle that is bright coppery-green in colour and covered in small, easily dislodged scales that give it a powdery appearance. The elytra are marked with a series of fine dark lines that run along their length. The legs are stout and antennae thick and segmented. **SEASON** Adult occurs Apr–Aug. **HABITAT** Verges, wasteland, woodland, gardens and other areas where nettles and low-growing plants occur. **HABITS** Popularly referred to as the Nettle Weevil or Green Nettle Weevil due to it often being observed feeding on Common Nettle. Despite its name, it feeds on a variety of low-growing plants. **STATUS** Widespread and fairly common.

Hazel Weevil ■ *Curculio nucum* Length 6mm

DESCRIPTION A small beetle with a distinctively long and thin, proboscis-like snout or rostrum. Oval in shape, the thorax and elytra are mottled light reddish-brown in colour. The legs are robust and brown, the eyes black. Antennae are thread-like and positioned halfway along the snout. **SEASON** Adult occurs Apr–Jun. **HABITAT** Hedgerows, deciduous woodland and gardens. **HABITS** Adults feed on hazelnuts and leaves. The female lays a single egg inside a maturing hazelnut within which the larvae feed and develop, overwintering in the fallen nut and emerging in the spring as adults. **STATUS** Widespread and locally common where Hazel occurs. Scarce in the extreme north.

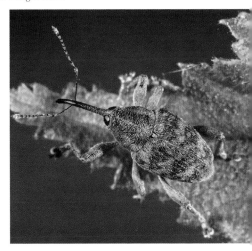

Acorn Weevil ■ *Curculio glandium* Length 5-8mm

DESCRIPTION Similar to the Hazel Weevil (above), the adult has a distinctively long and thin, proboscis-like rostrum. Oval in shape, the thorax and elytra are covered in scales and are mottled brown in colour. The legs are robust and brown, the eyes black. Antennae are thread-like and positioned along the rostrum. **SEASON** Adult occurs Apr–Jul. **HABITAT** Hedgerows, deciduous woodland and gardens. **HABITS** Adults feed on acorns and oak leaves. The female bores into an acorn with her rostrum and lays a single egg inside within which the larvae feed and develop, emerging in the spring. **STATUS** Common and widespread in suitable habitats. In Britain, common only in the south.

WEBSITES

UK Moths - http://ukmoths.org.uk/
An illustrated guide to the moths of Great Britain and Ireland.

UK Butterflies - http://www.ukbutterflies.co.uk
The website provides information on all of the butterfly species found in the British Isles.

Atropos - http://www.atropos.info
A resource for visitors to find out about latest sightings, how to get involved in recording in their local area and to share information with other enthusiasts.

Butterfly Conservation - http://butterfly-conservation.org/
A group whose aims are to conserve threatened butterflies and moths in the UK and elsewhere as well as to undertake and promote the scientific study of butterflies, moths and methods needed to conserve their habitats.

Buglife - http://www.buglife.org.uk/
Buglife aims to provide the scientific knowledge and information to everyone who needs or wants to know about how we can better look after our bugs.

British Dragonfly Society - http://www.british-dragonflies.org.uk/
The Society has two main inter-linked areas of interest, dragonfly recording and dragonfly conservation.

Wildlife Trusts - http://www.wildlifetrusts.org/
There are 47 individual Wildlife Trusts covering the whole of the UK and the Isle of Man and Alderney. Together, The Wildlife Trusts are the UK's largest, people-powered environmental organization working for nature's recovery on land and at sea.

Natural History Museum - http://www.nhm.ac.uk/nature-online/
The Natural History Museum is a leading scientific research institution, a major cultural attraction and recorder of life on Earth.

SUGGESTED READING

Ball, S. & Morris, R. (2013). *Britain's Hoverflies: An Introduction to the Hoverflies of Britain*. Princeton University Press.
Chinery, M. (2005). *Complete Guide to British Insects*. HarperCollins.
Brooks, S. *Field Guide to the Dragonflies and Damselflies of Great Britain and Ireland*. British Wildlife Publishing.
Greenhaigh, M. and Ovenden, D. (2007). *Freshwater life of Britain and Northern Europe*. HarperCollins.
Sterry, P. (2010). *Complete Guide to British Garden Wildlife*. HarperCollins.
Taylor, M. (2010). *A Naturalist's Guide to Garden Wildlife*. John Beaufoy Publishing.
Townsend, M. & Waring, P. & Lewington, R. (2007). *Concise Guide to the Moths of Great Britain and Ireland*. British Wildlife Publishing.
Thomas, J. & Lewington, R. (2014). *The Butterflies of Britain and Ireland*. British Wildlife Publishing.

▪ INDEX ▪

157